SALINITY AND TIDES IN ALLUVIAL ESTUARIES

SALINITY AND TIDES IN ALLUVIAL ESTUARIES

by

HUBERT H.G. SAVENIJE

Delft University of Technology
2600 GA Delft, The Netherlands

ELSEVIER

Amsterdam • Boston • Heidelberg • London • New York • Oxford
Paris • San Diego • San Francisco • Singapore • Sydney • Tokyo

ELSEVIER BV
Radarweg 29, PO Box 211
1000 AE Amsterdam
The Netherlands

ELSEVIER Inc
525 B Street, Suite 1900
San Diego, CA 92101-4495
USA

ELSEVIER Ltd
The Boulevard, Langford Lane
Kidlington, Oxford OX5 1GB
UK

ELSEVIER Ltd
84 Theobalds Road
London WC1X 8RR
UK

First edition 2005

ISBN-13: 978-0-444-52107-1 (Hardbound) 978-0-444-52108-8 (Paperback)

∞ The paper used in this publication meets the requirements of ANSI/NISO Z39.48-1992 (Permanence of Paper).

05 06 07 08 09 10 10 9 8 7 6 5 4 3 2 1

CONTENTS

PREFACE

Since the publication of "*Mixing in Inland and Coastal Waters*" by Fischer et al. (1979) and "*Estuaries, a Physical Introduction*" by Dyer (1973, revised in 1997), no comprehensive textbook has been published on salinity and tides in estuaries. Although considerable knowledge has been gained since, and several articles have been written on the subject, this has not yet resulted in a book that combines this knowledge within one consistent theoretical framework.

An estuary is the transition between a river and a sea. There are two main drivers: the river that discharges fresh water into the estuary and the sea that fills the estuary with salty water, on the rhythm of the tide. The salinity of the estuary water is the result of the balance between two opposing fluxes: a tide-driven saltwater flux that penetrates the estuary through mixing, and a freshwater flux that flushes the saltwater back. Both fluxes strongly depend on the topography. The salt water flux, because the amount of water entering the estuary depends on the surface area of the estuary, and the fresh water flux, because the cross-sectional area of the estuary determines the efficiency of the fresh water flow to push back the salt.

So, the topography is crucial. It provides the most important boundary condition for tidal hydraulics, mixing, and salt intrusion. One of the innovations of this book is that it makes use of the natural topography of alluvial estuaries, throughout. The natural topography of alluvial estuaries is one with converging banks following an exponential function. Both the width and the cross-sectional area obey an exponential function. Moreover, in coastal plain estuaries, the depth is constant and there is no bottom slope. Estuaries in coastal areas with a strong relief are generally too short for this type of estuary to develop. They form a special category of alluvial estuaries where standing waves occur and where the depth decreases in upstream direction. These estuaries are described e.g. by Wright et al. (1973) and Prandle (2003). The length scale of the convergence rate is a key parameter for understanding tidal and mixing processes. This book systematically integrates these natural topographies with tidal movement, mixing, and salt intrusion.

Mixing in estuaries is driven by both the tide and the density gradient. The density gradient induces vertical mixing; the tide mainly horizontal mixing through tidal trapping and residual circulation (and to a minor extent turbulent mixing). It is recognized that residual circulation is a dominant mechanism, particularly near the mouth of the estuary, but it is poorly understood. Several mixing mechanisms have been well documented in the literature, such as the vertical density-driven circulation and turbulence-driven mixing, but, as yet, no theory exists for residual circulation. In this book, it will be demonstrated that residual circulation is strongly related to the topography, and particularly to the width, which in alluvial estuaries is predictable. As a result, the book will provide an integrated mixing theory and a practical computational approach for the prediction of salt intrusion and tidal propagation in alluvial estuaries.

In 1992, I wrote my PhD thesis on the subject of this book. Since then, I have written many articles on alluvial estuaries, ranging from a predictive method for salt intrusion to developing analytical equations for tidal damping, tidal wave propagation, and the influence of fresh water discharge thereon.

This book presents a review of the state of the art, a comprehensive theoretical background on estuary shape, tides, mixing, and salt intrusion in alluvial estuaries and will present ample case illustrations from all over the world. It will provide tools with which human interference in estuary dynamics can be described and predicted, resulting from, for instance: upstream fresh water abstraction, dredging or sea-level rise. In describing the interactions between tide, topography, water quality and river discharge, it will provide useful information for hydraulic engineers, morphologists, ecologists, and people concerned with water quality in alluvial estuaries.

The book builds on the theory presented in *Environmental Fluid Dynamics* by Jörg Imberger, describing the fundamentals of fluid mechanics of free surface flow. This book continues on Imberger's book within the geometrical setting of an alluvial estuary, without unnecessary duplication. For details on the fundamentals of fluid mechanics the reader should refer to this standard work. The book that is lying in front of you can both be used as a text book for graduate students and as a reference work for researchers and practitioners.

In writing this book I received strong support from Professor Murugesu Sivapalan and Professor Jörg Imberger of the Centre for Water Research of the University of Western Australia, who invited me to spend my sabbatical leave with them, supported by a Gledden Fellowship. I am very grateful to both of them and all the staff of this magnificent institute for their support and hospitality. My wife and I look back on a very pleasant and fruitful time in Perth.

Hubert H.G. Savenije
Delft

NOTATION

a	Cross-sectional convergence length [L]
A	Cross-sectional area [L^2]
$\hat{A}$	Amplitude of the cross section [L^2]
A_f	Cross-sectional area of the flood channel [L^2]
A_0	Cross-sectional area at the estuary mouth [L^2]
b	Convergence length of the stream width [L]
B	Stream width [L]
B_L	Width at which there are no ebb and flood channels [L]
B_S	Storage width [L]
B_0	Width at the estuary mouth [L]
c	Wave celerity [L/T]
c_i	x-dependent coefficient equal to the ratio between dispersion coefficient and the fresh water velocity [-]
c_0	Classical wave celerity [L/T]
C	Convergence term (Chapter 3) [-]
C	Coefficient of Chézy [$L^{0.5}/T$]
d_s	Sediment transport factor [$L^{(2-n)}T^{(n-1)}$]
D	Longitudinal dispersion [L^2/T]
D_{ef}	Longitudinal dispersion due to residual circulation in ebb and flood channels [L^2/T]
D_i	Damping term (Chapter 3) [-]
D_g	Longitudinal dispersion due to gravitational circulation [L^2/T]
D_{SS}	Steady state dispersion [L^2/T]
D_t	Tide driven dispersion [L^2/T]
D_{50}	Diameter of the bed material that is exceeded by 50% of the sample weight [L]
e_p	Pumping efficiency: relative difference of the tidal velocity amplitude between flood and ebb currents [-]
E	Tidal excursion [L]
E_d	Dissipated energy [ML^2T^{-2}]
E_m	Potential energy per tidal period [ML^2T^{-2}]
E_t	Kinetic energy per tidal period [ML^2T^{-2}]
f	Friction factor [-]

f'	Adjusted friction factor [-]
f''	Adjusted friction factor [-]
f_D	Darcy-Weisbach friction factor [-]
F	Force per unit mass [$MT^{-2}L^{-2}$]
F	Mass flux [M/T]
F_e	Salt flux in the ebb channel [M/T]
F_f	Salt flux in the flood channel [M/T]
$\mathbf{F}$	Froude number (υ/c) [-]
$\mathbf{F_d}$	Densimetric Froude number (υ/c) [-]
g	Acceleration due to gravity [L/T^2]
h	Stream depth [L]
$\bar{h}$	Tidal average stream depth [L]
h_0	Constant tidal average stream depth [L]
H	Tidal range [L]
H'	Slack tidal range [L]
i	Side slope [-]
I	Water level slope [-]
I_r	Density slope [-]
I_b	Bottom slope [-]
k	Tidal wave number [L^{-1}]
k_s	Coefficient of proportionality in Lacey's equation [$T^{0.5}L^{-0.5}$]
K	Manning's coefficient [-]
K	Dimensionless Van den Burgh's coefficient [-]
L_T	Length of the tidal intrusion [L]
L	Salt intrusion length [L]
L_{ef}	Length of a ebb-flood channel loop [L]
M	Moment per unit volume driven by gravitational circulation [$M\ L^{-1}\ T^{-2}$]
N	Canter Cremers Estuary number [-]
N_E	Wave-type number [-]
N_R	Estuarine Richardson number [-]
N_P	Prandle's estuary number [-]
O	Surface area of the estuary [L]
P	Wetted perimeter [L]
P_t	Flood volume [L^3]
q	Coefficient of the advective term (Chapter 5) [L/T]
Q	Discharge [L^3/T]
Q_b	Bankfull discharge [L^3/T]
Q_f	The freshet or fresh water flushing [L^3/T]
Q_p	Tidal peak discharge [L^3/T]
Q_r	The river discharge [L^3/T]
Q_s	Sediment discharge [L^3/T]
Q_t	Tidal discharge [L^3/T]
Q_0	Dry season fresh water discharge [L^3/T]
r	Net rainfall (the difference between rainfall and evaporation) [L/T]

r_S	Storage width ratio [-]
r_0	Net rainfall rate during the dry season [L/T]
R	Friction term (in Chapter 2) [L/T^2]
R	Friction term (in Chapter 3) [-]
R'	Resistance term [T^{-1}]
R_L	Lorentz linearized friction factor [T^{-1}]
R_s	Source term [L^2/T]
s	Salinity [M/L^3]
S	Distance travelled by a water particle (in Chapter 2) [L]
S	Bottom slope term (in Chapter 3) [-]
S	Steady state salinity [M/L^3]
S_f	Fresh water salinity [M/L^3]
t	Time [T]
T	Tidal period [T]
T_f	Particle travel time [T]
T_K	Time scale for the system response [T]
T_Q	Time scale of the discharge reduction [T]
T_S	Time scale for the system response (steady state) [T]
U	Mean cross-sectional flow velocity [L/T]
U_b	Velocity of the bankfull discharge [L^3/T]
U_f	Velocity of the fresh water discharge [L/T]
V	Velocity of the moving particle [L/T]
x	Distance [L]
y	Dimensionless tidal range H/H_0 [-]
z	Vertical ordinate [L]
Z	Water level [L]
Z_b	Bottom elevation [L]
α	Tidal Froude number (in Chapter 3) [-]
α	Mixing coefficient (in Chapter 5) [L^{-1}]
α'	Adjusted tidal Froude number [-]
α_S	Shape factor [-]
α_L	Coefficient for the length of an ebb–flood interaction cell [-]
β	Damping scale (in Chapter 3) [L]
β	Dispersion reduction rate (in Chapter 5) [-]
β_L	Coefficient for the length of an ebb–flood interaction cell [-]
Δ	Relative density [-]
δ_H	Damping rate of tidal range [L^{-1}]
δ_U	Damping rate of tidal velocity amplitude [L^{-1}]
ε	Phase lag between HW and HWS, or LW and LWS [-]
η	Tidal amplitude [L]
ϑ	Damping factor [-]
λ	Length of the tidal wave [L]
μ	Amplitude growth factor [-]
ν	Ratio of the tide-driven dispersion to total dispersion [-]

ξ	Dimensionless argument (Lagrangean variable) [-]
ρ	Density of the water [ML^{-3}]
ς	Dimensionless salinity
υ	Tidal velocity amplitude [L/T]
φ	Angle or phase lag [-]
Φ	Harmonic function of the tidal velocity
Ψ	Harmonic function of the water level
ω	Angular velocity [T^{-1}]

Abbreviations:

LWS	Low water slack
HWS	High water slack
LW	Low water
HW	High water
TA	Tidal average

1

Introduction: description and classification of alluvial estuaries

What is special about alluvial estuaries and what makes them different from non-alluvial and man-made estuaries? This chapter presents a classification of estuaries and related types of salt intrusion. Subsequently it focuses on the characteristics, peculiarities, and the resulting behavior of alluvial estuaries.

The shape of an alluvial estuary is similar all over the world. The width reduces in upstream direction as an exponential function. In coastal plain estuaries, there is no significant bottom slope, but in estuaries with strong relief, the depth may decrease exponentially. As a result in both types of alluvial estuaries, the cross-sectional area varies exponentially, and so does the flood volume, also called 'tidal prism' (the volume of water that enters the estuary on the tide). Morphological equilibrium and minimum stream power lie at the cause of this typical shape. There is a similarity between the exponential reduction of the width of an estuary (in upstream direction) and the exponential increase of the drainage area of a catchment (in downstream direction) described in Rodriguez-Iturbe and Rinaldo (2001).

The constant depth and exponentially varying width correspond with the shape of an 'ideal estuary' as described by Pillsbury (1939) and Langbein (1963), a topography that can be observed in coastal plain estuaries all over the world. The shape of an alluvial estuary is characterized by the ratio of the depth of flow (h) to the convergence length (b), the length scale of the exponential function. Although there is both empirical evidence (examples of 15 estuaries will be provided) and mathematical proof of this fact, the acceptance of this phenomenon is still low among scientists who are used to work with a different schematization of the topography (most authors assume a bottom slope and many use a constant width). A classification of estuaries will be presented taking into account tidal range, river flow, type of salt intrusion, estuary topography, and relief.

The topography is key to estuary processes. The fact that water flows as it does is strongly influenced by the medium through which it flows. In principle, the non-linear hydraulic equations (the St. Venant equations) can demonstrate irregular and unpredictable behavior, but in alluvial estuaries, we seldom observe this. Instead we observe a number of surprisingly simple 'laws':

1. The tidal excursion (the distance a water particle travels during a tidal cycle) is constant along a coastal plain estuary. This is related to the morphological equilibrium (Savenije, 1989).
2. There are simple analytical relations for estuary topography (h/b), wave celerity, and phase lag, that can be derived from the equation for conservation of mass (Savenije, 1992a, 1993b).
3. We also see that tidal amplification obeys a simple linear function, whereas tidal damping is partly linear and partly exponential, based on the equation for conservation of momentum (Savenije, 1998, 2001a; Horrevoets et al., 2004). Although this equation is more complex than 'Green's Law,' it is still surprisingly simple.
4. The propagation of the tidal wave is influenced by tidal damping (and *vice versa*). This interaction can also be described by a simple analytical equation. (Savenije and Veling, 2005).
5. The phase lag between the moment of high water and the subsequent moment of slack, when the current changes direction, is a key parameter in tidal hydraulics, often disregarded.
6. We observe that salt intrusion is well mixed or partially mixed at the time when it matters. In tidal estuaries, the salt wedge, which most people think is the dominant salt intrusion mechanism, either does not occur at all, or only occurs during high river floods, when nobody is worried about the salt intrusion but rather about flood protection (Savenije, 1992b).
7. Mixing of salt and fresh water, although a complex process that results from many different mixing mechanisms, can be described by a surprisingly simple formula, originally coined by Van den Burg (1972).
8. Salt intrusion can be described by an analytical equation that can be applied to new situations with a minimum of calibration. In fact the equation is predictable in that it can be applied outside the range of calibration, e.g. to analyze the effect of river discharge, interventions in the estuary by dredging, sea-level rise, etc. (Savenije, 1993a,c).

These phenomena will be briefly introduced in this chapter, but duly described and derived from the basic equations in the subsequent chapters.

1.1 IMPORTANCE OF ESTUARIES TO MANKIND

An estuary is the transition between two distinct water bodies: a river and a sea. One of the few things that a river and a sea have in common is that they contain water and hence, provide an aquatic environment. The differences, however, are more abundant: a river transports and does not retain water, whereas a sea primarily stores water; river water is fresh, whereas sea water is saline; a river has more or less parallel banks and a bottom slope in the direction of flow, whereas a sea is virtually unlimited and has no bottom slope in the direction of flow; in a sea the tidal waves—perpendicular to the coast—are primarily standing waves, whereas in a river, the flood waves are primarily progressive (for definitions, see Chapter 2). Certainly, this list is not exhaustive.

An estuary, on the other hand, has characteristics of both a river and a sea (see Table 1.1). Typical riverine characteristics of an estuary are that it has banks, flowing water, sediment transport, occasional floods, and—in the upper parts—fresh water. Typical marine characteristics are the presence of tides, and saline water. But the most typical feature of an estuary is that it is the transition between a river and a sea, with its own hydraulic, morphologic, and biologic characteristics such as: tidal waves of a mixed type, a funnel shape, and a brackish environment, quite different from other water bodies. Rivers carry nutrients to the nutrient-poor oceanic environment. The estuary is the region where these two environments interact, serving as a crucial feeding and breeding ground for many life forms. Because of these typical characteristics and the related unique habitats, the estuarine environment plays an important role in the life cycle of numerous species. The flora and fauna of estuaries are extremely rich and the area of influence through the migration of species is large. The importance of estuaries to the global environment therefore is not easily overestimated.

To man, estuaries have always been important, both as a source of food and as a transport link between river and sea. In addition, lands bordering estuaries generally have excellent potential for agriculture: soils are fertile, the land is flat, and fresh water is, in principle, available. Hence the most densely populated areas of the world are situated in coastal areas near estuaries. However, estuaries are also fragile. An estuary is a sediment sink, accumulating sediments stemming both from the river and the sea. As a result, the residence time of pollutants, attached to the sediments, is high. Given the high value of the estuarine ecosystem, the intense human use of the coastal zone and the high susceptibility

Table 1.1 Characteristics of an estuary compared to a river and a sea

	Sea	Estuary	River
Shape	Basin	Funnel	Prismatic
Main hydraulic function	Storage	Storage and transport	Transport of water and sediments
Flow direction	No dominant direction	Dual direction	Single downstream direction
Bottom slope	No slope	No slope	Downward slope
Salinity	Salt	Brackish	Fresh
Wave type	Standing	Mixed	Progressive
Ecosystem	Nutrient poor, marine	High biomass productivity, high biodiversity	Nutrient rich, riverine

to pollution, it is imperative that estuarine environments are protected. In some areas, interference of man in the water resource system has led to serious deterioration of invaluable ecosystems. As a consequence of the growing concern towards these developments, management for sustainable use of the estuary water resources is receiving increasing attention.

This book aims at supplying insight into the dominant hydraulic and hydrologic processes that determine the estuarine environment, and in doing so, it provides a tool for improving the sustainable management of the water resources of estuaries. With the formulae derived in this book, one can determine and predict the impact of interventions in the estuary system (such as dredging or fresh water withdrawal) on the tidal range, the tidal propagation, the mixing processes, and the salt intrusion. They can be used to determine the amount of fresh water that needs to be released to counterbalance saltwater intrusion and to compute the longitudinal distribution of the salinity as a function of geometric, hydraulic, and hydrological parameters.

Knowledge of the physical phenomena that determine the process of tidal flow, tidal mixing, and salt intrusion are important as they influence the environment of an estuary and its water resource potential in many ways. Interventions in the estuary topography can have drastic impacts on the hydraulic behavior of an estuary, which may cause dramatic, and often irreversible, ecological changes. Upstream developments, such as river regulation by dams and fresh water withdrawal, have a direct impact on the salinity distribution in the estuary. Changes in the salinity distribution impact on water quality, water utilization and agricultural development in the coastal area, and the aquatic environment in general.

The importance of estuaries to mankind cannot be easily overstated. We are witnessing a period of rapid development and change, at a pace unprecedented in history. Our almost unstoppable drive for further development lies at the heart of most of our problems. The global processes of development have unleashed forces that are difficult to contain. At the same time, our knowledge of the natural system is growing fast and the technology at our disposal and the related computational capacity is increasing by the day. This book's contribution to the sustainable development of coastal zones is that it aims at enhancing our insight and understanding of the processes at play in the estuarine environment, in the hope that the next generation of scientists, engineers, and water managers will find the inspiration and the means to use the estuary resources wisely and conserve what is most fragile and valuable.

1.2 CLASSIFICATION OF ESTUARIES

The main driving forces affecting the character of an estuary are:

- the tide, which is generated by the interaction of the earth with the sun and the moon and acts as a periodic function consisting of multiple components, depending on the frequencies of the earth's rotation, the orbits of the earth and the moon, and other planetary processes at longer time scales. The dominant periods are in the order of 12.3 (semi-diurnal) and 24 (diurnal) hours. The tide is the main supplier of energy and salt water to the estuary system. The tide is responsible for the harmonic pumping of water into and out of the estuary with an erosive power that is neutralized only if the banks converge at an exponential rate.

- the waves, which have a stochastic nature, depending on meteorological conditions. Waves can have a dominant influence on the formation of the estuary mouth. The amount of energy supplied by waves, particularly during extreme events can be substantial, however, in contrast to the tidal energy, which dissipates along the entire estuary axis, its energy dissipation is concentrated near the mouth of the estuary.
- the river discharge, which provides fresh water and sediments to the estuary system. The river sediments deposit on entering the estuary, as soon as the banks of the river widen. These sediments can only be transported downstream by the residual downstream energy of the flow, which is a combination of the harmonic tidal flow and the downward river flow. If the banks are parallel, the downstream transport process is faster. Hence estuaries with a high river discharge (particularly during floods) tend to have modest bank convergence (near parallel banks).
- lateral (littoral) sediment transport along the coast. The transport of sediments parallel to the coast can be responsible for the formation of spits and bars. If the littoral transport is strong in relation to the erosive power of the tide and of the river discharge, then this can lead to estuaries that are (temporarily) closed off from the sea by an ephemeral bar. Such an estuary is called a 'blind' estuary.
- the density difference, which is responsible for a residual inward current along the bottom, transporting marine sediments into the estuary. This process can lead to mud or sand bar formation at the upstream limit of the salt intrusion.

These forces lead to different shapes of estuaries, each with a different aquatic environment. There are several ways of classifying these estuaries, according to: a) shape, b) tidal influence, c) river influence, d) geology, and e) salinity.

(a) Classification based on shape

In estuaries, the following characteristic shapes can be distinguished:

- prismatic. The banks of the estuary are parallel. This is a type of estuary that only exists in a man-made environment where the banks are artificially fixed. Examples are shipping channels that are regularly dredged and where the banks are stabilized. In an estuary where the flood volume reduces in upstream direction, and consequently the flow velocity amplitude as well, no morphological stability is possible. A constant cross section can only be maintained through dredging, for example, the Rotterdam Waterway.

- delta. A near prismatic estuary where the tidal influence is small compared to the amount of river water feeding the delta. Deltas occur in seas with a relatively small tidal range and on rivers with a high sediment load (e.g. the Mississippi, the Nile, the Mekong).
- funnel or trumpet shape. The banks converge in upstream direction. This is the natural shape of an alluvial estuary, where the tidal energy is equally spread along the estuary axis (e.g. the Maputo, the Pungué, the Schelde).
- drowned valleys and Fjords. Fjords are the result of glaciers that eroded the underlying rock, after which the valley was submerged by sea level rise. Drowned river valleys (also called by the Portuguese name Ria) stem from the irregular topography of a watershed drowned by sea level rise, where the feeding rivers carry too little sediment to keep up with the sea level rise (e.g. the Bay of Sydney). The latter type generally has irregular banks with several side channels and embayments.
- bays. These are semi-enclosed bodies that do not have a significant input from a river. The distinction between a bay and a drowned valley is often not easy to make.

A good description of several of these estuaries is provided by Dyer (1997, pp. 7–12).

(b) Classification on the basis of tidal influence

The most general classification used is by Davies (1964), who distinguishes micro-, meso-, macro-, and hyper tidal estuaries on the basis of the tidal range (see Section 2.2.1). This is assuming that the tidal range is a good indicator of the amount of tidal energy that is dissipated in the estuary, which is responsible for the erosive power of the tide and hence, the shape of the estuary. Nichols and Biggs (1985) introduced the term synchronous estuary where the amount of tidal energy per unit width is constant along the estuary axis and the tidal amplitude is constant. This term is erroneous since synchronicity suggests that water levels in the estuary are reached at the same time (which happened in estuaries experiencing a standing wave), but that is not what the authors meant. What they meant is the following:

- an ideal estuary. An estuary where, as the tidal wave travels upstream, the amount of energy per unit width lost by friction is exactly equal to the amount of energy gained by convergence of the banks. In an ideal estuary, the tidal range is constant along the estuary axis.
- an amplified estuary. An estuary where the tidal range increases in upstream direction because convergence is stronger than friction. Clearly this process cannot continue indefinitely, implying that at some point along the estuary, the friction should become more pronounced leading to a reduction of tidal amplification and subsequently to tidal damping. The process of damping is enhanced by the river discharge, which increases the friction and reduces bank convergence.

- a damped estuary. An estuary where friction outweighs bank convergence. Tidal damping occurs in estuaries with a long convergence length or in drowned river values with a narrow opening.

(c) Classification based on river influence

If we classify on the basis of river influence, we can distinguish two extreme cases, the riverine and the marine estuary:

- the riverine estuary. This estuary is dominated by the river flow, both in terms of discharge and sediment supply. The water is fresh and it behaves like a river: parallel banks, regular bank overtopping if not protected by dikes, a sandy bottom, and sandy banks. The tide propagates as a progressive wave.
- the marine estuary. This estuary is dominated by the sea. The water is completely saline. There is no significant fresh water and sediment input from the land side. The banks are often muddy. The ecosystem is primarily marine. The tide propagates as a standing wave.

There is obviously a clear link between this classification and the classification based on shape. Riverine estuaries are prismatic or delta estuaries, as discussed above, whereas marine estuaries are bays.

(d) Classification based on geology

- Fixed bed estuary. An estuary with a fixed bed is a remnant of a different geological era. Alluvial estuaries are very young in geological terms, whereas fixed bed estuaries stem from an older geological period. Fjords are the remnants of glaciers that were active during the ice ages, and Rias are remnants of drowned river basins where the river does not generate sufficient water and sediment to keep up with the rate of sea level rise.
- Alluvial in a coastal plain. An alluvial estuary consists of sediments that have been deposited by the two water bodies that feed it: the river and the sea. Within these sediments, the estuary has shaped its own bed in a way that the energy available for erosion and deposition is equally spread along the estuary, resulting in a situation of minimum stream power. The coastal plain is long enough for the alluvial estuary to develop fully.
- Alluvial on a short coastal plain, in a submerged valley (Ria) or Fjord. These are estuaries that are alluvial in the sense that the water flows in its own sediments, but they have not yet reached a stage of morphological equilibrium. The geological formation process (e.g. sea level rise or tectonic dip) is too fast for the sedimentation to keep up. These estuaries have not been able to develop an equilibrium length. Some estuaries are too short for an equilibrium length to develop, because the topography is too steep, resulting in standing waves, for instance the Ord river described by Wright et al. (1973), others are too long, resulting in a complex system of lagoons and drowned river

channels (e.g. the Ria de Aveiro in Portugal or the Swan river in Western Australia).

(e) Classification based on salinity

- Positive or normal estuaries. A positive or normal estuary is an estuary where the salinity decreases gradually in upstream direction, from sea salinity to river water salinity. These estuaries are the dominant type in temperate and wet tropical climates. Because there is a significant river input, these estuaries are generally alluvial.
- Negative or hypersaline estuaries, however, have a salinity that increases upstream due to the fact that evaporation exceeds rainfall and the amount of fresh water input from the river is too small to compensate for the difference. These estuaries occur in arid and semi-arid climates and are characterized by the occurrence of salt flats (salinas). Because of the low fresh water discharge, and subsequently the low sediment input, these estuaries are often not fully alluvial but more commonly, Rias. Hypersaline estuaries have a very peculiar ecosystem often dominated by pink algae. Ample attention to these estuaries is given in Chapter 4.

1.3 ESTUARY NUMBERS

In the classification of estuaries we saw that the two dominant drivers of an estuary, that influence their character, are the tide and the river discharge. The simplest dimensionless number that characterizes this ratio is the estuary number N, which in the Dutch literature is called the Canter-Cremers[1] number, equal to the ratio between the amount of fresh and saline water entering the estuary during a tidal period. The first volume is the product of the fresh water discharge Q_f and the tidal period T, the latter is the flood volume P_t, also called the tidal prism:

$$N = \frac{Q_f T}{P_t} \tag{1.1}$$

Another important estuary number is the Estuarine Richardson[2] number, which is defined as the ratio of potential energy provided to the estuary by the river

[1] J.J. Canter-Cremers (1879–1925) was a Dutch hydraulic engineer working for the Ministry of Public Works on the propagation of storm surges and tides in river branches. He laid the foundation for the Dutch research on tidal propagation in estuaries, essential for the prestigious 'Delta-Works' constructed in the branches of the Rhine, Schelde, and Meuse delta during the twentieth century.

[2] Lewis Fry Richardson (1881–1953) was a British scientist with a very broad background. He held degrees in physics, mathematics, chemistry, biology, and psychology. His field of work was very broad. His earliest publications are on meteorology while his latest work is on the psychology of war and peace. In hydraulic engineering, he is known for his theory on fluid convective stability, leading to the Richardson number, and for his law on turbulent diffusion.

discharge through buoyancy of fresh water and the kinetic energy provided by the tide during a tidal period:

$$N_R = \frac{\Delta\rho}{\rho} \frac{gh}{\upsilon^2} \frac{Q_f T}{P_t} \tag{1.2}$$

This estuary number accounts for more driving factors than the Canter-Cremers number. It incorporates the effect of the relative density difference between fresh water and seawater and of the Froude number, which is the ratio between the amplitude of the tidal velocity υ and the celerity of a finite amplitude wave ($c_0 = \sqrt{(gh)}$):

$$F = \frac{\upsilon}{c_0} \tag{1.3}$$

If the Estuarine Richardson number is high, there is enough potential energy available in the river discharge to maintain a sharp interface and subsequently, stratification occurs; if it is low, there is enough kinetic energy available in the tidal currents to mix the river water with saline water and the estuary is well mixed. These estuary numbers, which are key in the classification of estuaries, are derived and presented in more detail in Chapter 2.

In Table 1.2 a combined overview of the different estuary types is presented, together with their main characteristics related to tide, river influence, geology,

Table 1.2 Estuary classification in relation to the aspects: tide, river influence, geology, salinity, and estuary number

Shape	Tidal wave type	River influence	Geology	Salinity	Estuarine Richardson Number
Bay	Standing wave	No river discharge	–	Sea salinity	Zero
Ria	Mixed wave	Small river discharge	Drowned drainage system	High salinity, often hypersaline	Small
Fjord	Mixed wave	Modest river discharge	Drowned glacier valley	Partially mixed to stratified	High
Funnel	Mixed wave; large tidal range	Seasonal discharge	Alluvial in coastal plain	Well mixed	Low
Delta	Mixed wave; small tidal range	Seasonal discharge	Alluvial in coastal plain	Partially mixed	Medium
Infinite prismatic channel	Progressive wave	Seasonal discharge	Man-made	Partially mixed to stratified	High

salinity, and estuary number. We see that the tidal wave type and the type of the salinity intrusion are very different in these estuaries. Some of the characteristics are not always obvious. For instance in Fjords, which generally do not drain large catchments (otherwise they would already have become alluvial) and hence do not have a very high river discharge, stratification can still occur. This is because they generally are very deep and, as a result, have a relatively low tidal velocity amplitude (and tidal excursion), leading to a high Estuarine Richardson number. In prismatic channels, stratification can also occur if there is sufficient river water forced through the channel.

1.4 ALLUVIAL ESTUARIES AND THEIR CHARACTERISTICS

Alluvial estuaries are estuaries that have movable beds, consisting of sediments of both riverine and marine origin, in which there is a measurable influence of fresh water inflow. The water moving in the estuary can either erode the estuary bed (by deepening or widening) or it can deposit sediments and—in doing so—make the estuary narrower or shallower. Hence, the shape of an alluvial estuary is directly related to the hydraulics of the estuary, or as Wright et al. (1973) put it: 'the simultaneous co-adjustment of both process and form has yielded an equilibrium situation.' This equilibrium is a dynamic equilibrium between deposition and erosion, where at some point in time, erosion is dominant and at another point in time, deposition. This happens at different time scales: the short inter-tidal time scale where the tidal velocity accelerates and decelerates within the tidal period, and at a seasonal or annual time scale as a result of river discharge variation. The residual current resulting from the river discharge gradually, and slowly, transports sediments downstream, but during floods it deposits sediments because the transport capacity of flow reduces as the estuary widens and the flow decelerates.

So in an alluvial estuary, the water movement depends on the topography and the topography in turn depends on the erosive power of the hydraulics. The interdependence between hydraulics and topography is important because it permits us to derive hydraulic information from the estuary shape and to derive geometric information from the hydraulics.

Pethick (1984) elaborating on the differences between alluvial estuaries and rivers stated that 'in its simplest form' the major difference is 'that the tidal flow of the estuary, unlike the flow of a river is not going anywhere.' Although in a river water flows because upstream rainfall generated a discharge that seeks a way downstream, the water in an estuary flows just because the estuary is there. If the estuary were not there, the water would not flow. It requires an opening in the coastline, with a storage area behind it, for the estuary water to flow. Hence the flow in an estuary is completely governed by the shape of the estuary.

Table 1.3 Interaction between shape and flow in estuaries and rivers

	Shape determines discharge	Shape does not determine discharge
Discharge determines shape	Alluvial estuaries	Alluvial rivers
Discharge does not determine shape	Fjords and Rias	Canals and non-alluvial rivers

The amount of flow entering and leaving the estuary depends on the channel size. In a river, the discharge does not depend on the channel size, it depends on the rainfall that has fallen in the upstream catchment. In summary, the tidal discharge in estuaries depends on channel size, whereas river discharge does not.

In an estuary with fixed beds (e.g. Fjords), the estuary shape is not affected by the flow (although the flow is affected by the shape). In such estuaries, no sediment is supplied from a feeding river and, if there is no continental shelf, there is also no supply of sediment from the sea. In estuaries with fixed beds, there is no interdependence between hydraulics and shape and hence, it is not possible to derive universal relations between shape and hydraulics in these estuaries (see Table 1.3).

1.4.1 The shape of alluvial estuaries

Alluvial estuaries have converging banks that can be described by an exponential function. The equations describing estuary shape and hydraulics will be presented in Chapter 2. Here we shall discuss the shape in more general terms. Figure 1.1 shows a view from the top of a typical estuary. The origin of the longitudinal axis of the estuary is located at the mouth, which is the point where the estuary meets the ocean. The exact position of this point is often difficult to determine, but it can generally be found by connecting the adjacent shorelines by a fluid line. A good definition to use is the point where the primarily one-dimensional flow pattern of the estuary changes into two-dimensional flow, or the point where the ebb and flood currents that fill and empty the estuary are fanning out into the ocean.

In Figure 1.1, we see two important horizontal length scales of the estuary: the width B, which is a function of x, and the tidal excursion E, which in alluvial estuaries appears to be constant. The tidal excursion is the distance that a water particle travels on the tide. It moves inland during the flood tide and moves out again during ebb. The tidal excursion is in fact the horizontal tidal range, which is the integral over time of the tidal velocity between the two moments of slack: low water slack (LWS) and high water slack (HWS). The typical length of the

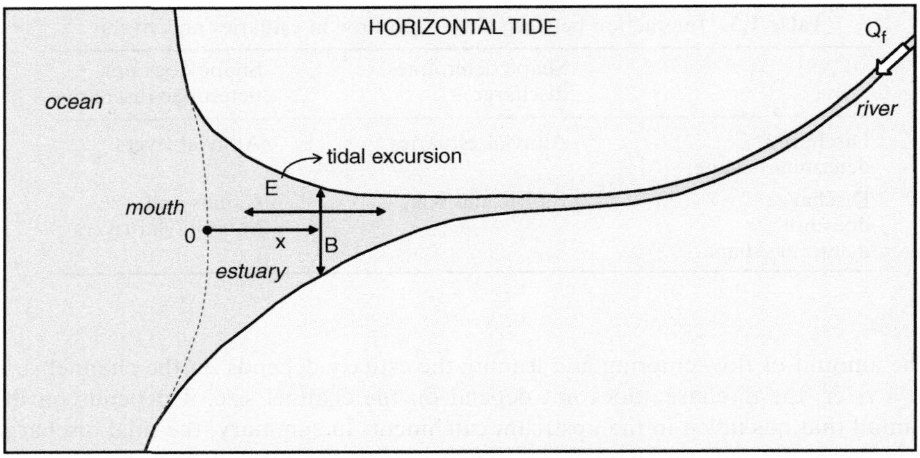

Figure 1.1 Definition sketch of an estuary: view from the top.

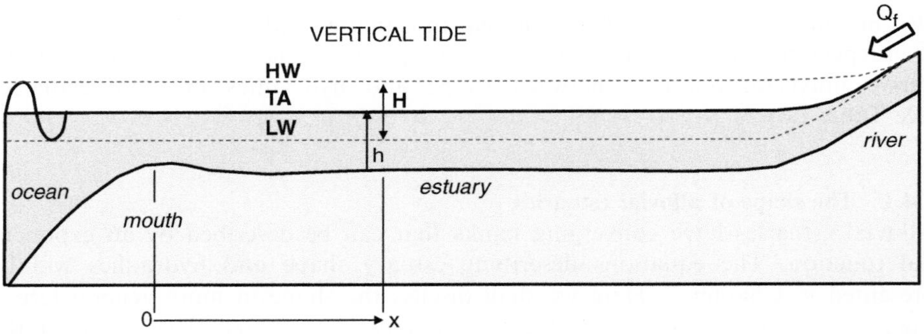

Figure 1.2 Definition sketch of an estuary: longitudinal cross section.

tidal excursion for a diurnal tide is 10 km. In the widest part of the estuary, the ratio of E/B can be small, particularly in strongly funnel-shaped estuaries, but in the upper part of the estuary, this ratio is generally large.

In Figure 1.2, we see a longitudinal cross-cut of the estuary. What strikes us is that the estuary bottom is horizontal. It is in agreement with Pethick's (1984) observation that the tidal flow is not going anywhere. It is logical that where there is no resultant direction of flow, there is also no resultant bottom slope. In reality, the depth of the estuary fluctuates as the flow meanders through the channel, deep in the bends and shallow in the crossings, but on an average, in alluvial estuaries, there is no bottom slope, contrary to what most people believe. The bottom slope only starts where the estuary gradually changes into the river and the river discharge Q_f becomes dominant over the tidal currents. This phenomenon is discussed in Section 2.2. Also in the longitudinal cross section, we see two important vertical length scales. The (vertical) tidal range H and the depth of flow h. The ratio of tidal range to depth is an important tidal characteristic. If this ratio is small, then the non-linear hydraulic equations may be linearized and relatively simple solutions of the hydraulic equations are possible. If this ratio is large, however, second-order effects in the non-linear hydraulic equations become dominant and simple solutions are no longer possible.

Figure 1.2 shows two envelopes: one for high water (HW) and one for low water (LW). At all times, the water levels in the estuary remain between these envelopes. The mean water level (TA) is essentially horizontal and links up with the backwater curve of the river. We see that the two envelopes also converge to the water level of the river, which is sloped. In the river, the water level slope is the same for high and low flows, but during floods, the river influence reaches further. Being a schematic picture, these processes are somewhat exaggerated.

1.4.2 Dominant mixing processes

There are two main drivers for mixing in estuaries: the density difference and the tide. The tide drives the flow and the density difference drives an imbalance in hydrostatic pressure that is discussed in Section 2.1.4. These two driving mechanisms generate four main mixing mechanisms:

1. Turbulent mixing or shear mixing. In a river cross section, there is a balance between the tidal driving force (gravity) and friction. Although gravity works on all water particles, the friction only works along the bottom and banks of the estuary. The friction force is transported to all other particles through a shear stress that is transferred by turbulence. As a result, the flow velocity in a cross section is generally highest at the largest distance from the bottom and the banks. The turbulence that is associated with this shear stress also causes mixing, much as in a regular river. Turbulent mixing has been intensively studied and subsequently described by hydraulic engineers, but in estuaries, this type of mixing is not so important. The following three mechanisms are much more efficient in mixing salt and fresh water.
2. Gravitational mixing. Gravitational mixing stems from the fact that the hydrostatic pressure on the sea-side and on the river-side do not attain equilibrium. Because sea water is denser than river water, the pressure on the ocean-side would be higher at equal depth than on the river-side. As a result, the water level at the limit of the salt intrusion is slightly higher than at sea (about 10 cm if the estuary is 8 m deep). Although on an average the hydrostatic forces cancel out, the pressures are not equal over the depth. Near the surface, the resultant pressure is directed towards the sea, while near the bottom, it is directed upstream. As a result, there is a residual circulation that carries relatively saline water upstream along the bottom and relatively fresh water

downstream along the surface. The vertical salinity gradient that stems from that is an important cause of saline and fresh water mixing, particularly in the part of the estuary where the salinity gradient is large. Also this phenomenon has been amply studied, both in laboratories and in the field, and analytical equations have been derived to describe this mixing process.

3. 'Trapping.' Tidal trapping is the result of the irregularity of the banks of an estuary. If there are tidal inlets or tidal flats, then the water that fills these water bodies on the incoming tide is generally released into water with different densities. There is a phase lag between the filling and emptying of tidal flats and the flow in the main channel, which results in density differences. In irregular channels, tidal trapping can be an important mechanism, particularly in large estuaries with tidal flats. The tidal excursion is the dominant mixing length scale of tidal trapping, since this is the maximum distance over which a water particle can travel. Also this mechanism is more important with a larger salinity gradient.
4. 'Tidal pumping' or mixing by residual currents. This mechanism is least studied, yet it is a very important one (as also observed by Fischer et al., 1979). Particularly near the mouth of a wide estuary, this is the dominant mechanism. The most important difference between the other two dominant mixing mechanisms (gravitational circulation and trapping) is that it does not depend on the salinity gradient, but that it is proportional to the estuary width. Not many researchers have dedicated analytical effort to this mixing mechanism. Rather they concentrated their efforts on the middle reach of an estuary where gravitational circulation is dominant.

In Chapter 4, these mixing mechanisms are dealt with in detail and a predictive theory is presented that allows these mixing mechanisms to be integrated into one analytical equation: the modified Van den Burgh equation, developed by Savenije (1992a) after Van den Burgh (1972).

1.4.3 How the tide propagates

In an alluvial estuary, the tide propagates as a wave with both a standing and a progressive character. A progressive wave is much like the wave that the bow of a boat generates and that we feel when we are swimming in the water. At the maximum elevation, the water moves in the direction of the wave; at the lowest elevation of the wave the velocity is backward, in the direction opposite to the direction of the wave; at the average water level, the movement is vertical. The highest flow velocity occurs at maximum elevation. Of a progressive wave, the phase lag between the water elevation and the velocity is zero. Flood waves in rivers are also progressive waves. At maximum elevation of the flood, the discharge is also at its maximum; at the lowest elevation, the discharge is at its minimum. In general, these waves occur in frictionless channels of infinite length and with a constant cross section. It is not so important that the flow is completely frictionless, but for a progressive wave to occur, the channel should have a constant cross section and be very long.

A standing wave is different. A standing wave is like a wave we can create in a tub by rocking it. After we have rocked the tub, the water in the tub continues to rock back and forth. In the tub, the maximum and minimum water levels are reached all at the same time. The water just swings back and forth like a swing. At the extremes, the velocity is zero and changes direction. Here we see a phase lag between elevation and velocity of $\pi/2$. Standing waves occur in harbors, bays and, in general, semi-enclosed basins that are connected to the sea.

Alluvial estuaries are none of the above. In alluvial estuaries, waves occur of the mixed type, where the phase lag is between zero and $\pi/2$. In estuaries with a semi-diurnal tide, the flow velocity slacks about 40 min after HW or LW.

If we make a longitudinal cross section along the estuary, we can draw the envelopes of HW and LW, as in Figure 1.2. If we would observe the instantaneous water levels in the estuary, then we would obtain an instantaneous tidal wave as shown schematically in Figure 1.3. This wave is contained between the envelopes. A full tidal wave seldom fits within the length of an estuary (except in very long estuaries like the Gambia), and the slope of the river is somewhat exaggerated. In Figure 1.4, some more instantaneous tidal waves are drawn between the HW and LW envelopes. The figure suggests that the tidal range is constant as we move

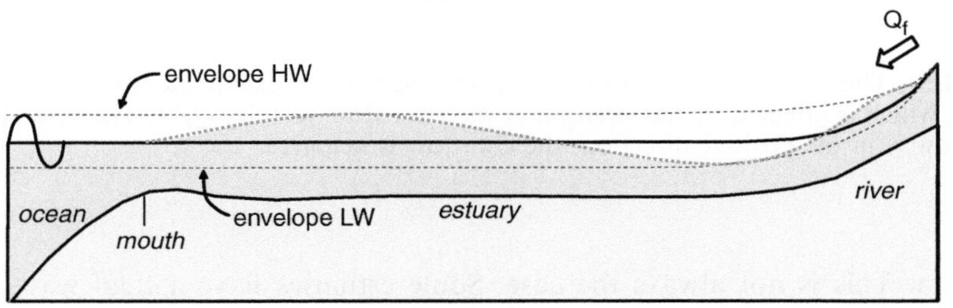

Figure 1.3 Longitudinal cross section showing an instantaneous tidal wave.

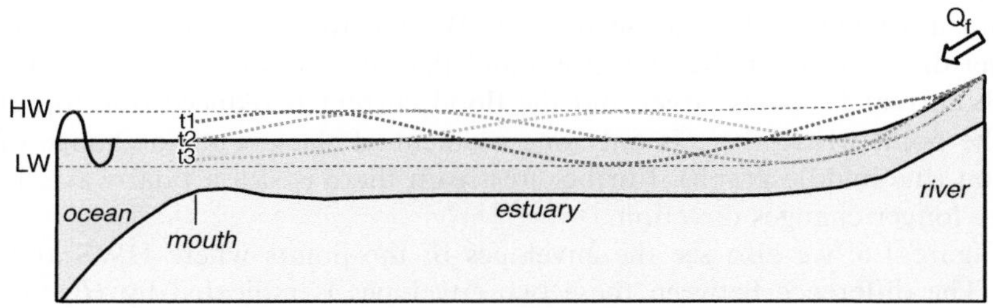

Figure 1.4 Instantaneous water levels contained between envelopes of HW and LW.

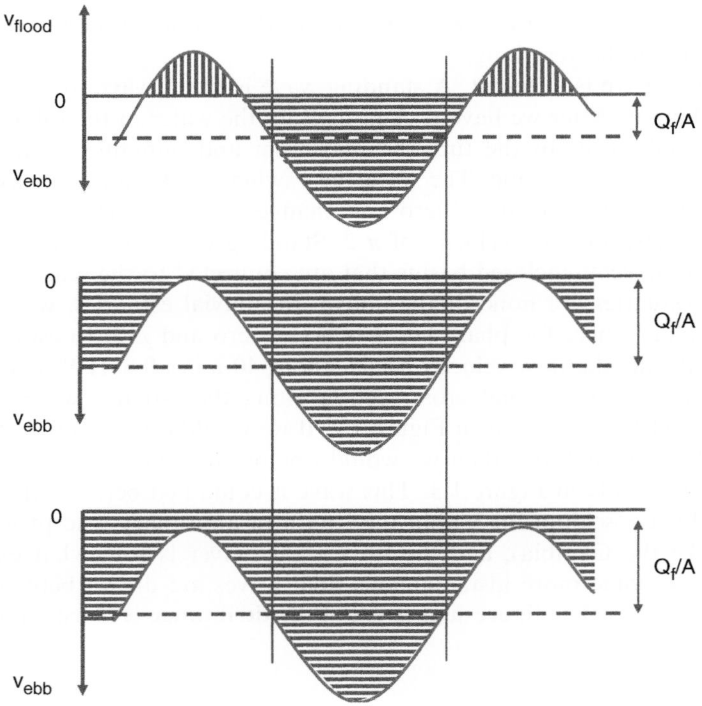

Figure 1.5 The influence of river discharge. On the vertical axis is the flow velocity of the current, which is affected by the strength of the river flow velocity Q_f/A. The velocity of the flood flow is in the positive direction; the ebb flow is negative.

upstream. This is not always the case. Some estuaries have a tidal wave that is damped, while others experience tidal amplification. In the latter case, the tidal range is only damped as the river influence becomes dominant over the tide (e.g. in the Schelde Estuary).

As the river discharge gains influence, the velocities are modified. Figure 1.5 provides a schematic illustration of that. We see that there is superposition of the river discharge over the tidal flows and that as a result of the river discharge the ebb current becomes larger and the flood current is reduced (the top graph), until the point where there is only one moment of slack, but no change of flow direction (the middle graph). Further upstream there is still a tidal wave, but the flow no longer changes direction.

In Figure 1.6, we also see the envelopes of the points where HWS and LWS occur. The difference between these two envelopes is indicated by H', the slack tidal range. The figure also shows the point P, where the two slacks coincide and upstream from which the flow no longer changes direction. The dashed part

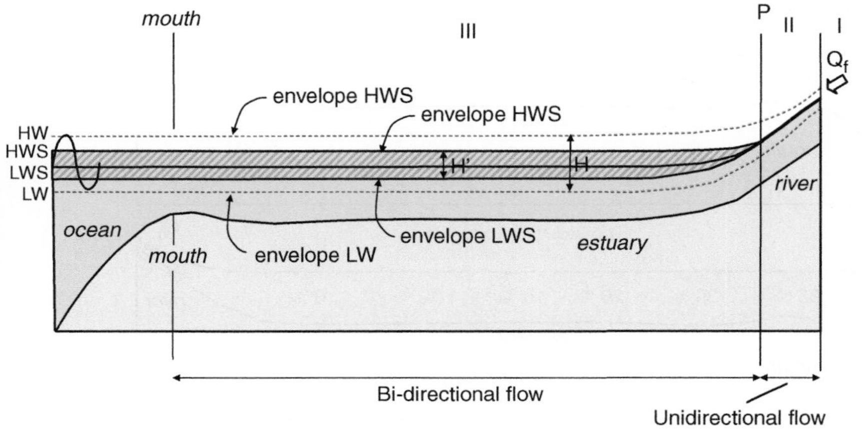

Figure 1.6 The tidal prism contained between the envelopes of HWS and LWS.

between the envelopes for HWS and LWS is equal to the tidal prism P_t, as we shall see in Section 2.3.2.

1.4.4 How the salt intrudes

Types of salt intrusion

The salt intrusion mechanism is generally divided into three types:

a. the stratified type, or the saline wedge type,
b. the partially mixed type,
c. the well-mixed type.

Figure 1.7 illustrates the three types. A stratified estuary occurs when the fresh water discharge in an estuary is large as compared to the tidal flows (a large Canter-Cremers number and a large Richardson number). A well-mixed estuary occurs when the fresh water discharge is small compared to the tidal flows (a small Canter-Cremers number, and a small Estuarine Richardson number). Figure 1.8 shows the related vertical salinity gradient: a sudden increase of the salinity over the depth in a stratified estuary, a smooth gradient in a partially mixed estuary, and the absence of a gradient in a well-mixed estuary.

What is seldom seen in figures like this is the relative length of the salt intrusion. For instance, the sketch presented in Fischer et al. (1979; Figure 7.1) gives a wrong impression: it suggests that in stratified, partially mixed, and well-mixed estuaries the intrusion length reaches equally far and that it can occur at the same river discharge. It also suggests a bottom slope. In contrast, Figure 1.7 clearly indicates that a saline wedge only occurs close to the mouth and during a period

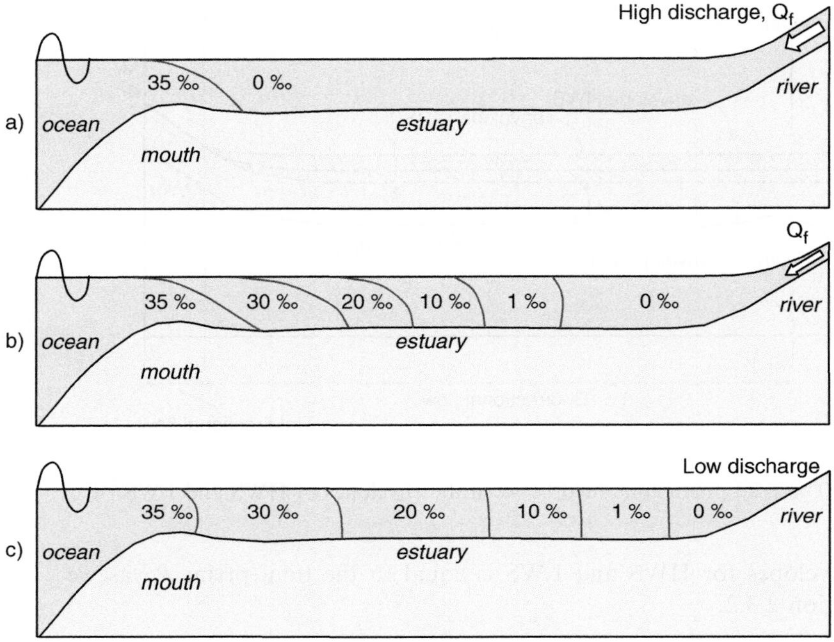

Figure 1.7 Longitudinal distribution of the salinity for a stratified estuary (a), a partially mixed estuary (b), and a well-mixed estuary (c).

of high river discharge (when the Canter-Cremers number is large). The well-mixed estuary occurs during low flow, resulting in a much deeper salt intrusion.

During the dry season, when water availability is lowest, water requirements are highest. When the problems of maintaining an acceptable water quality are most pronounced, the salt intrusion is generally of the mixed type. As the water consumption increases, the salt intrusion becomes even more mixed as the Canter-Cremers number reduces further. Especially if one aims at making optimum use of the water resources available, the critical case to be considered for design is the well-mixed type when salt intrusion is at its maximum. Hence, the well-mixed type is the most interesting case for the water resources manager. In alluvial estuaries, the saline wedge only occurs during periods of high river discharge; a time at which we are hardly interested in salt intrusion, but rather in flood protection (in man-made estuaries with a constant width, this may be different due to the associated high Estuarine Richardson number). The relatively high attention given to stratified salt intrusion by hydraulic engineers, hence, is more related with their professional interest than with a societal need.

The distinction between the partially mixed and the well-mixed type is arbitrary. The salt intrusion is generally regarded as well mixed when the stratification (the difference between the salinity at the water surface and the salinity near the

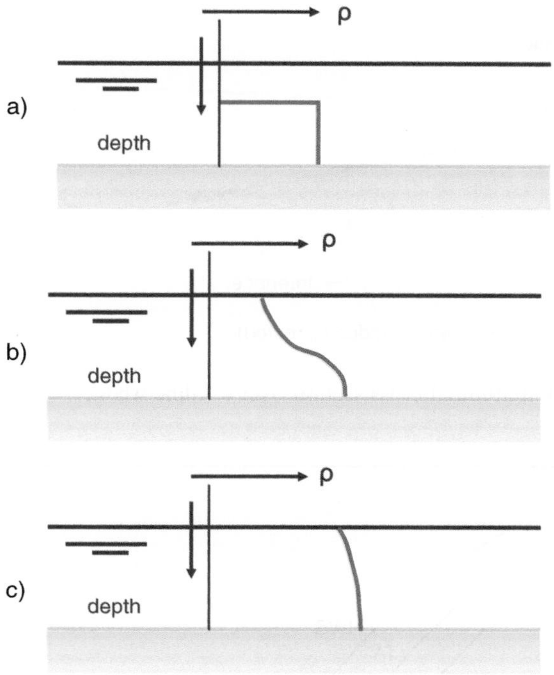

igure 1.8 Variation of the salinity over the depth in a stratified (a), partially mixed (b), nd well-mixed estuary (c).

ottom divided by their average) is less than 10 percent. This is no objective iterion. Near the sea, there is more stratification than further upstream; and oreover the value of 10 percent is arbitrary. In practice, however, the value of) percent is not so important. Until a stratification of 20–30 percent is reached, serious drawbacks have been encountered in applying well-mixed theory.

Figure 1.9 shows a longitudinal cross-cut over a saline wedge. We see that the ater level increases slightly in upstream direction as a result of the density adient. As a result, there is a resultant downstream fresh water flow in the upper yer and a resultant upstream flow in the salt layer. We can see that there is sharp interface but that along the interface, there is entrainment of saline water y the fresh water that flows over it. In fact, the slope of the saline wedge is aintained by a shear stress exercised by the fresh water discharge. This shear ress and the related turbulent mixing is responsible for the downward salt ansport that counterbalances the upstream salt transport over the bottom.

Figure 1.10 shows the longitudinal distribution of the salinity along a well-mixed tuary. We see that there are different intrusion lines for HWS (the maximum lt intrusion), LWS (the minimum salt intrusion), and for the tidal average (TA)

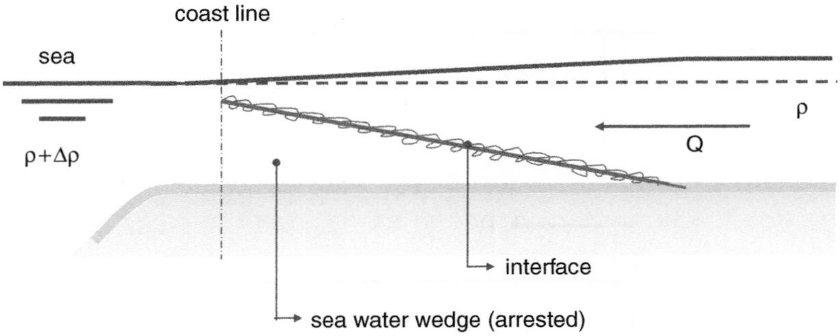

Figure 1.9 A longitudinal cross section over a saline wedge.

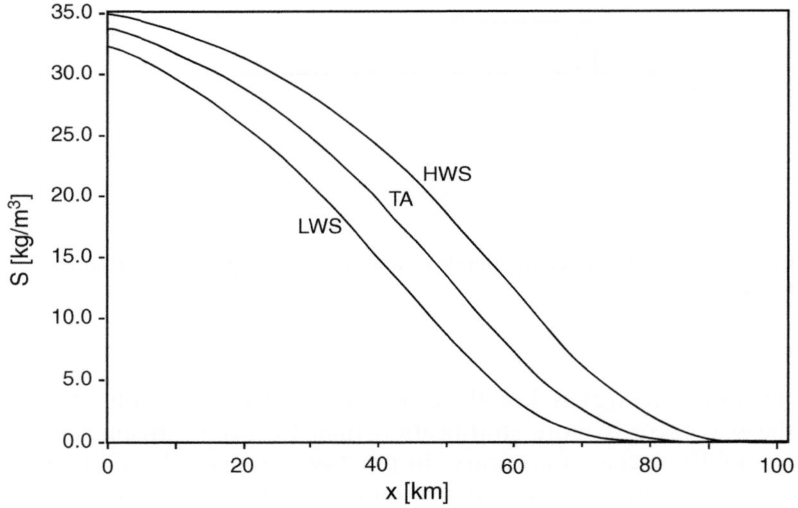

Figure 1.10 The longitudinal salinity distribution in a well-mixed estuary at high water slack (HWS), low water slack (LWS) and tidal average (TA) condition.

situation. The horizontal distance between the HWS and LWS curve is the tidal excursion. We see that the salinity moves up and down the estuary following the water particles that travel between HWS and LWS. The horizontal translation of the curves demonstrates a constant tidal excursion along the estuary.

Figure 1.11, finally, shows the different shapes of well-mixed salt intrusion curves that can be distinguished.

1. The recession shape, which occurs in narrow estuaries with a near-prismatic shape and a high river discharge. This type is more common not only in

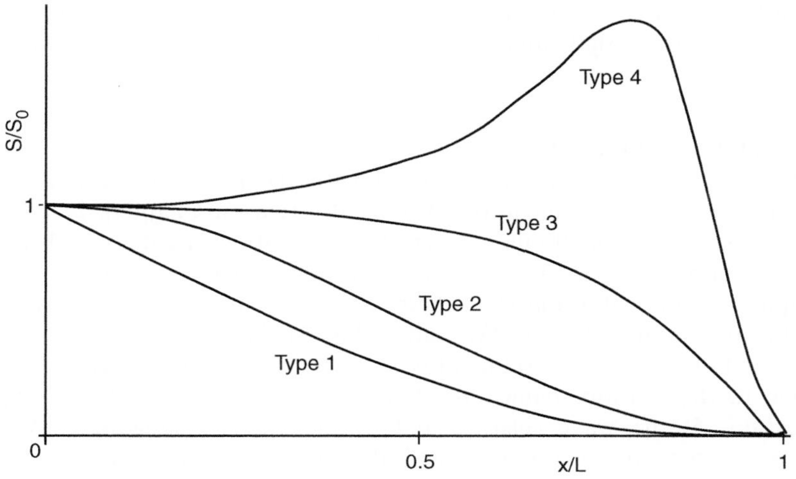

Figure 1.11 Different shapes of well-mixed salt intrusion curves.

deltas and shipping channels, but also is dominant in riverine estuaries such as: the Corantijn (Suriname), the Chao Phya (Thailand), the Limpopo (Mozambique), the Solo (Indonesia), and the Rotterdam Waterway (The Netherlands).

2. The bell shape, which occurs in estuaries that have a trumpet shape, i.e. a long convergence length in the upstream part, but a short convergence length near the mouth. There are several estuaries that have a stronger widening near the estuary mouth, possibly related with littoral processes. Examples are: the Mae Klong (Thailand), the Incomati (Mozambique), and the Maputo (Mozambique).
3. The dome shape, which occurs in strong funnel-shaped estuaries (with a short convergence length). Examples are: the Schelde (The Netherlands), the Thames (England), the Pungué (Mozambique), and the Delaware (USA).
4. The humpback shape, which is a negative or hypersaline estuary, mentioned above. The driver for hypersalinity is evaporation. If evaporation exceeds rainfall and fresh water inflow, an estuary can become hypersaline. Estuaries with dome-shaped salt intrusion are most susceptible to hypersalinity. Hypersalinity often occurs in Rias in semi-arid areas, which have large water bodies compared to the amount of fresh water input. Examples are the Saloum (Senegal) and the Casamance (Senegal).

We see that the first three shapes are very much determined by the topography of the estuary. Topography is a key driver of both tidal hydraulics and mixing in estuaries, and as a result, of salt intrusion. Therefore the interaction of estuary processes with topography is a key feature in this book.

The different types of salt intrusion curves are discussed in more detail in Chapter 4 where we shall analyze the physical processes that generate these shapes.

1.5 WHAT WILL FOLLOW

The following chapters will deal in detail with the issues of tidal hydraulics, mixing processes, and salt intrusion in alluvial estuaries. Chapters 2 and 3 deal with the theory of tidal hydraulics. A set of new equations are derived that follow from the combination of the general hydraulic equations with the typical topography of alluvial estuaries. These equations are analytical expressions for: 1) the phase lag between HW and HWS, 2) the relation between estuary geometry and the tidal range, 3) the ratio between the Froude number and the tidal amplitude to depth ratio, named the 'scaling equation,' 4) tidal damping and amplification, 5) the wave celerity, and 6) the channel roughness. All these equations are general versions of 'classical' equations commonly used to date, which have in common that they consider the effect of friction and channel convergence, as well as the fact that tidal waves are neither progressive waves nor standing waves, but waves of a mixed character with a phase lag between zero and $\pi/2$.

Chapter 4 subsequently provides a theory for mixing in alluvial estuaries, which builds upon the work carried out till date, but which introduces a theory on one of the most important mixing mechanisms of funnel-shaped estuaries: residual circulation by 'tidal pumping.' It will be shown that Van den Burgh's method is a very convenient approach to combine into one equation the effect of the three main mixing mechanisms: tidal pumping, trapping, and gravitational circulation.

Based on the two previous chapters, Chapter 5 presents a one-dimensional steady state and non-steady state model for salt intrusion in estuaries. The resulting equations are surprisingly simple and yield very good results in a wide range of estuaries studied worldwide. The method is predictive in that it can be extrapolated to discharge regimes outside the range of calibration and that it can be applied to new estuaries without the explicit need for further calibration. The unsteady model allows the application to estuaries that do not reach an equilibrium situation in the dry season, or that may become hypersaline. This makes the model a very useful tool to assess the consequences of upstream withdrawals and particularly to assess the risk of estuaries becoming hypersaline.

All in all, a complete theory of tidal hydraulics, tidal mixing, and salt intrusion is presented that can assist both the researcher and the water manager to study the behavior of alluvial estuaries in a rapidly changing environment.

2

Tide and estuary shape

As with all open channel flow, tidal flow in estuaries can be described by the St. Venant equations: a set of two non-linear partial differential equations that govern the movement of water through a medium. What makes tidal flow in alluvial estuaries different from other hydraulic phenomena is the medium through which the water flows. As we saw in the previous chapter, in coastal plains this medium has a particular shape, similar to the shape of an ideal estuary. Although this knowledge is far from new, in practice only few people make use of it, probably because modern computational power allows us to make three-dimensional computations that no longer require geometric simplification. The mere application, however, of computer models without the knowledge and insight provided by the use of analytical equations, is often dangerous. Analytical solutions not only provide insight into the processes at play, more importantly, they provide a means for verification or falsification.

This chapter describes the hydraulic equations of alluvial estuaries where there is a close interaction between geometry and flow, mutually influencing each other in continuous feedback. As a result, a regular topography appears in which mathematical laws can be discerned that can be described by surprisingly simple analytical equations. In combining the conservation of mass and momentum equations with the topography of an alluvial estuary, a number of analytical equations are derived for: 1) tidal propagation, 2) tidal damping, 3) tidal amplification, 4) wave celerity, 5) phase lag, and 6) the influence of river flow on tidal damping. In Section 2.3, integration of the conservation of mass equation leads to the Geometry-Tide equation (a relation between topographical and tidal length scales) and an expression for wave celerity and phase lag (the Phase Lag equation). Combination of these two equations yields the Scaling equation. These equations are derived through Lagrangean[1] analysis, which is a mathematical approach more natural to estuary hydraulics and salt intrusion since the reference system moves

[1] Joseph-Louis Lagrange (1736–1813), a French mathematician and mathematical physicist was one of the greatest mathematicians of the eighteenth century. His work Mécanique Analytique (1788) was a mathematical masterpiece. Lagrange succeeded Euler as the director of the Berlin Academy. The term Lagrangean means: using a reference frame that moves with the water particle, or unit volume. Commonly the term Lagrangean is spelled with –ian, but this is wrong since the mathematicians name ends on an e, just like Shakespeare and Europe, both of which have adjectives on –ean.

with the water. This chapter deals with the relationship between hydraulic parameters and estuary shape. The next chapter looks into the tidal dynamics, presenting derivations for tidal damping (or amplification), tidal wave propagation, and their dependence on river discharge.

2.1 HYDRAULIC EQUATIONS

2.1.1 Basic equations

In alluvial estuaries, there is a dynamic equilibrium between erosion and sedimentation of sediments that are picked up, transported, and deposited by water. The water movement, in turn, strongly depends on the geometry it has created. This close interaction between the dynamics of water and sediment is an important characteristic of alluvial estuaries, as has been discussed in the previous chapter. The movement of water and sediment is generally described by a set of four one-dimensional equations: the conservation of momentum and mass for water, the conservation of mass for sediment, and an empirical formula that relates sediment transport to flow parameters (see e.g. Jansen et al., 1979):

$$\frac{\partial Q}{\partial t}+\alpha_S\frac{\partial(Q^2/A)}{\partial x}+gA\frac{\partial h}{\partial x}+gA\frac{\partial Z_b}{\partial x}+gA\frac{h}{2\rho}\frac{\partial\rho}{\partial x}+gA\frac{U|U|}{C^2h}=0 \tag{2.1}$$

$$r_S\frac{\partial A}{\partial t}+\frac{\partial Q}{\partial x}=R_s \tag{2.2}$$

$$B\frac{\partial Z_b}{\partial t}+\frac{\partial Q_s}{\partial x}=0 \tag{2.3}$$

$$Q_s=Bd_sU^n \tag{2.4}$$

where:

- $Q=Q(x,t)$ is the discharge in m^3/s;
- α_S is a shape factor (assumed constant) to account for the spatial variation of the flow velocity over the cross section ($\alpha_S>1$);
- $A=A(x,t)$ is the cross-sectional area of the flow in m^2;
- $h=h(x,t)$ is the mean cross-sectional depth of flow in m;
- $Z_b=Z_b(x,t)$ is the mean cross-sectional bottom elevation in m;
- g is the acceleration due to gravity in m/s^2;
- $\rho=\rho(x,t)$ is the density of the fluid in kg/m^3;
- $U=U(x,t)$ is the mean cross-sectional flow velocity in m/s;
- $C=C(x)$ is the coefficient of Chézy in $m^{0.5}/s$;
- $B=B(x,t)$ is the stream width of the channel in m;
- $B_S=B_S(x,t)$ is the storage width of the channel in m;

- $r_S = r_S(x)$ is the ratio of storage width B_S to stream width B ($r_S > 1$);
- R_s is a source term, accounts for rainfall, evaporation, or lateral inflow in m^2/s;
- $Q_s = Q_s(x,t)$ is the sediment discharge in terms of sediment volume (including pores) in m^3/s;
- n is an exponent;
- $d_s = d_s(x)$ is a parameter with the dimension $m^{(2-n)}s^{(n-1)}$ that depends on sediment characteristics and channel roughness.

Throughout this book, since the most important boundary condition lies at the estuary mouth, the positive x-direction chosen is the upstream direction with the origin at the sea or ocean boundary. The first two equations (derived and discussed in Imberger's book on Environmental Fluid Dynamics) are generally known as the St. Venant equations (named after A.J.C. Barré de Saint-Venant[2]).

The first equation, Equation 2.1, is the equation for conservation of momentum, derived from Newton's[3] second law of motion, stating that the acceleration of an object is equal to the balance of forces, in this case the component of gravity in the direction of flow and friction. The first term in Equation 2.1 is the Eulerian[4] acceleration term while the second term is the convective acceleration term. The coefficient α_S accounts for the shape of the channel. The more irregular a cross section and the more the variation in flow velocity over the cross section, the larger the α_S. It is larger than unity, but generally smaller than 2. In a regularly shaped, single channel, alluvial stream, α_S is usually close to unity (Jansen et al., 1979). In estuaries where there are no floodplains that discharge considerable parts of the flow, α_S is normally close to one.

The third, fourth, and fifth terms jointly represent gravity, exercised through the water pressure gradient. These terms are the gradient of the water depth, the bottom slope, and the density gradient. The density term is often disregarded, but it can play an important role in the brackish part of an estuary. For the derivation of this term, the assumption has been made that the density is merely a function of x and t and that there is no vertical salinity gradient (i.e. the estuary is well mixed). The fifth term will be discussed in detail in Section 2.1.4.

[2] In 1843, seven years after the death of Claude Navier (1785–1836), the Frenchman Adhémar Jean Claude Barré de Saint-Venant (1797–1886) re-derived Navier's equations for a viscous flow. In this article, he was the first to properly identify the coefficient of viscosity. He further identified viscous stresses acting within the fluid because of friction. George Stokes (1819–1903), like Saint-Venant, also derived the Navier–Stokes equations but he published the results two years after Saint-Venant (after J. J. O'Connor and E. F. Robertson).

[3] Isaac Newton (1643–1727) published his single greatest work, the Philosophiae Naturalis Principia Mathematica in 1686. It contains his famous laws of motion, and the law of universal gravitation.

[4] Leonhard Euler (1707–1783), a Swiss mathematician and student of Bernoulli, may be considered as the founding father of modern mathematics (introducing among other the exponential function, complex calculus, and the notation $f = f(x)$). His Introducio in Analysia Infinitorum (1748) provided the foundations of analysis. The term Eulerian is used for a reference frame that is fixed on the river or estuary bank, in contrast to a Lagrangean reference frame that moves with the water.

The last term of Equation 2.1 is the friction term, based on the formula of Chézy[5]. In this term, the depth h is used instead of the hydraulic radius. This assumption is justified if the estuary is wide in relation to its depth ($B \gg h$). In alluvial estuaries this is always the case. Since Chézy's coefficient is not independent of the depth, the formula of Manning[6] is considered more appropriate to describe the resistance term R:

$$R = g\frac{U|U|}{C^2 h} = g\frac{U|U|}{K^2 h^{2/3}} \tag{2.5}$$

where K is Manning's coefficient generally indicated by its inverse value n ($K = 1/n$).

The second St. Venant equation, Equation 2.2, is the conservation of mass equation, or the equation of continuity. In this equation, there is a balance between the first term, indicating the rate of increase of the volume over time, and the second term, indicating net inflow of water over the stretch considered. The sum of these terms should equal the source term, which accounts for lateral input of water from drainage, rainfall, or evaporation (negative). In this chapter, the source term can be disregarded where it relates to tidal hydraulics, since lateral inflow generally has a marginal influence on tidal parameters such as velocity and depth. In Chapter 4, however, the source term can play a key role in the salt balance equation, particularly when an estuary turns hypersaline.

In the first term of the second equation, the entire wet surface that stores water should be considered, not just the width of the stream where water flows. Hence, the need to take account of the ratio between storage width and stream width, $r_S > 1$.

The third equation is the conservation of mass equation of the sediment (or rather the conservation of volume). It represents the balance between sediment deposition over time and the increase of sediment transport over the reach considered. If the transport capacity increases erosion occurs, otherwise deposition. Erosion balances deposition when the sediment transport capacity is constant with x. The fourth equation is the sediment transport equation. It appears in several forms in the literature. The most widespread formula, which is well appreciated for its wide applicability in alluvial rivers as well as for its simplicity, is the formula

[5] In 1776, the French engineer Antoine de Chézy (1718–1798) published his well-known formula, which he had been using for some time, where the flow velocity is proportional to the root of the product of the hydraulic radius and the slope.

[6] Robert Manning (1816–1897), an Irish engineer, building on the work of De Chézy among others, published his well-known formula in 1891. Although he tried to make his coefficient dimensionless, he did not succeed. After the introduction of $\sqrt{g}$, there still remained a length to the power 1/6 to account for. This was done in 1923 by the Swiss hydraulic engineer Albert Strickler (1887–1963), who related the roughness to the 1/6th power of the ratio between effective roughness depth and water depth. As a result, the Manning formula is often called the Manning–Strickler formula.

of Engelund and Hansen (1967), where the exponent n equals 5 and the parameter d_s is defined by:

$$d_s = \frac{0.05}{D_{50}\Delta^2 C^3 \sqrt{g}} \tag{2.6}$$

where D_{50} is the diameter of the bed material that is exceeded by 50 percent of the sample by weight and Δ is the relative density of submerged sediment (generally $\Delta = (2600-1000)/1000 = 1.6$).

In addition, the following geometric relationships define A, r_S, and Q as:

$$A = hB \tag{2.7}$$

$$Q = UA \tag{2.8}$$

$$r_S = \frac{B_S}{B} \tag{2.9}$$

Finally there is an equation for the density gradient, which is not reproduced here. In Chapter 5, a relation will be presented that allows the determination of ρ as a function of space and time. For the following analysis it is assumed that the water density gradient is either known through measurements, or can be computed by an appropriate salt intrusion model. Assuming that α_S, r_S, ρ, C, g, n, Δ, D_{50}, and hence d_s are known, the list of dependent variables consists of the following seven parameters:

- the mean cross-sectional flow velocity $U(x,t)$
- the mean cross-sectional depth of flow $h(x,t)$
- the mean cross-sectional bottom elevation $Z_b(x,t)$
- the stream channel width $B(x,t)$
- the cross-sectional area $A(x,t)$
- the discharge $Q(x,t)$
- the sediment discharge $Q_s(x,t)$

Hence, there are six equations (Equations 2.1–2.4, 2.7, 2.8) with seven dependent variables. Consequently, one more equation is required, besides boundary conditions, to solve the set of equations for the seven dependent variables. The set of equations presented cannot be solved if there is no additional relation that relates geometric parameters to flow parameters. All conventional hydraulic models are based on the above equations, which can only be solved if the geometry of the channel (in particular the width) is fixed. With the present models, we are not yet able to predict what the shape of a channel will be when we provide a certain discharge at the upstream boundary of a freely erodable slope. Interesting new research documented by Rodriguez-Iturbe and Rinaldo (1997), using concepts such as self-organization, minimum stream power, and entropy, was undertaken to find this missing relation, but as yet the solution has not been found.

As a result, in computational hydraulics instead of a seventh equation, the width is imposed as a function of distance x and water level elevation ($Z = Z_b + h$). For a freely varying width, however, a 'seventh equation' is needed. In stable channel design, Lacey's formula is often proposed as the seventh equation. This equation is not dynamic, but it does provide an estimate of the equilibrium width of an alluvial channel.

2.1.2 The seventh equation

Although several efforts have been made to relate the width B to flow parameters, no unequivocal physically based method has yet been developed (to the disappointment of many researchers). For alluvial channels, Lacey, in 1930, formulated a theory based on an earlier work by Kennedy (1894) and Lindley (1919) which came to be known as 'regime' theory and which was based on the assumption that an alluvial channel adjusts its width, depth, and slope in accordance with the amount of water and the amount and kind of sediment supplied (Stevens and Nordin, 1987). Lacey's theory is almost entirely empirical and supplies simple power expressions that relate stream depth, width, slope, and velocity to the discharge. Regime theory has been relatively successful in India and Pakistan in the design of stable irrigation channels under natural regime. On the other hand, regime theory has been widely criticized mainly because of its lack of physical basis, its empirical character, and the scanty and incomplete database used for its derivation (Stevens and Nordin, 1987). Investigations by Stevens (1989) on stream width however indicated that, although there is still no satisfactory physical backing, there is also no reason to reject the empirical relationship between stream width and discharge (see also: Rodriguez-Iturbe and Rinaldo, 1997; pp. 12–15).

For his stream width formula, Lacey made use of the wetted perimeter P instead of the surface (or bottom) width B. The wetted perimeter is the length of the wetted cross-sectional profile over which shear stress is exercised, which is a better measure for the width in the friction term than the surface or bottom width. The wetted perimeter is somewhat larger than the width (in a rectangular profile $P = B + 2h$), but in alluvial streams where the width is generally much larger than the depth ($B \gg h$), the wetted perimeter is approximately equal to the stream width ($P \approx B$). Lacey found a surprisingly simple proportionality between the wetted perimeter and the root of the bankfull discharge:

$$B \approx P = k_S Q_b^{0.5} \tag{2.10}$$

where k_s, in metric units, equals 4.8 ($s^{0.5}m^{-0.5}$). Savenije (2003) showed that this coefficient of proportionality depends on the flow velocity at bankfull discharge U_b and the natural angle of repose φ of the bed material:

$$k_s = \sqrt{\frac{\pi^2}{2U_b \tan\varphi}} \tag{2.11}$$

The bankfull discharge is the discharge at which the river starts spilling over the natural levees. It is the discharge above which the river can deposit sediments on its banks. Regular overtopping is necessary for the river to maintain its bed. Leopold and Maddock (1953), who extended the regime concept to American rivers, confirmed that the width is proportional to the square root of the bankfull discharge. Blench (1952) arrived at the same conclusion and gave an expression for Lacey's coefficient k_s, which he related to the bed material and tractive force acting on the sides of the river bed. Later studies in American streams by Simons and Albertson (1960) showed similar results:

$$B = k_s {Q_b}^{0.51} \tag{2.12}$$

albeit that the exponent was slightly increased. The coefficient of proportionality k_s appeared to vary with the soil properties of the banks. The value of k_s varied between 3.1 for banks with coarse non-cohesive material to 6.3 for sandy banks (in metric units), which is in general agreement with Equation 2.11. The former value is lower than the latter because sandy banks are easier to erode. Lacey (1963), in the discussion of the article, maintained that an exponent of 0.5 is correct.

In estuaries, empirical studies of cross-sectional dimensions have yielded similar relations between tidal discharge and cross-sectional area. O'Brien (1931) presented a relationship between the cross-sectional area of the estuary mouth and the tidal flood volume P_t (the amount of sea water that enters the estuary on the flood tide) which, in its turn, is approximately proportional to the peak of the tidal discharge Q_p:

$$A \propto P_t^{0.85} \propto Q_p^{0.85} \tag{2.13}$$

In later studies, described by Bruun and Gerritsen (1960), other equations of the type of Equation 2.13 were derived based on the stable channel theory of Lane (1955) and Bretting (1958). Bretting's formula for estuaries reads:

$$A \propto Q_p^{0.9} \tag{2.14}$$

It can be shown that there is a striking resemblance between Equation 2.10 which applies to rivers, and Equations 2.13 or 2.14 which apply to estuaries.

If one assumes that the wetted perimeter P is approximately equal to the width B then application of Manning's formula to bankfull discharge Q_b yields:

$$Q_b = KBh^{1.7}\sqrt{I} \tag{2.15}$$

where I is the water level slope.

Substitution of the width B from Equation 2.10 in Manning's formula yields that the depth h is proportional to ${Q_b}^{0.3}$. Combination of this result with Equation 2.10 leads to:

$$A = hB \propto Q_b^{0.8} \tag{2.16}$$

which is close to Equation 2.14.

Savenije (2003) who considered bankfull flow as a singularity where Manning's equation no longer applies (because the water slope is forced by the overtopping levees and not by the balance between friction and gravity) found the exponents for B and h both to be equal to 0.5, leading to a direct proportionality between A and Q_b, and a bankfull velocity U_b that is independent of Q. This result is also close to Equation 2.14. In fact, Bruun and Gerritsen (1960) showed that in the tidal inlets between the islands along the Dutch coast, there was a direct proportionality

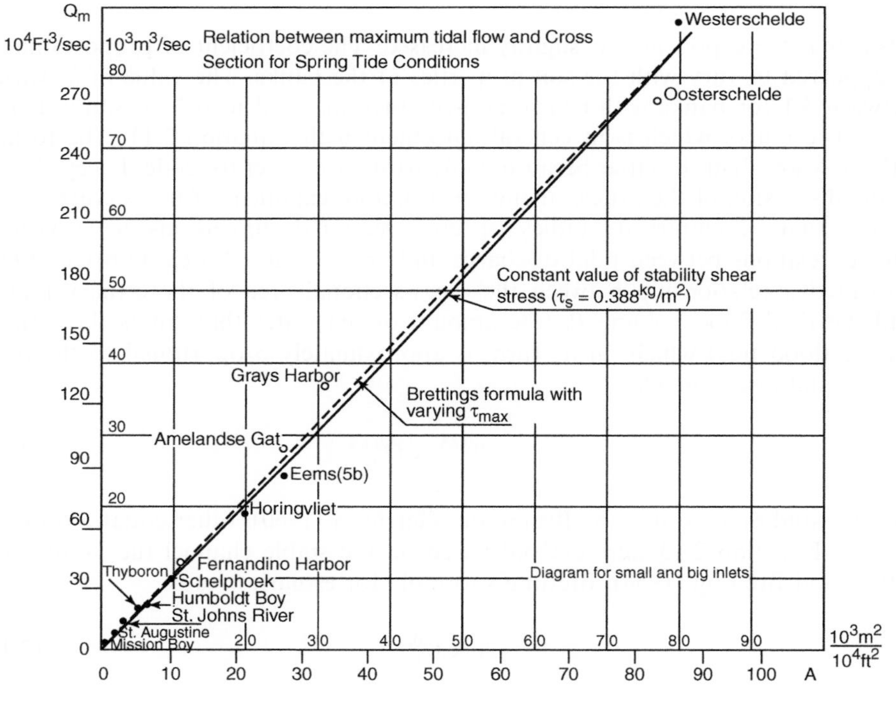

Figure 2.1 Relationship between peak tidal discharge Q_p and cross-sectional area of the tidal inlet A as reported by Bruun and Gerritsen (1960).

between Q_p and A at a rate of a tidal peak velocity $v = 1$ m/s (see Figure 2.1). Although Bruun and Gerritsen use an exponent of 0.9, an exponent of 1 with a peak velocity of 1 m/s is as feasible (see Figure 2.1). However, the fact that the exponent of Bruun and Gerritsen (0.9) lies in between 1 and 0.8 (between a value for bankfull and within-bank discharge, respectively) is an indication that in estuaries bankfull discharge (where the banks just overtop), is not always achieved.

The good correspondence between these equations for both estuaries and rivers suggests that estuaries do not substantially differ from alluvial rivers in terms of morphology. The main difference being, that in a river the bankfull discharge determines the channel shape, whereas in an estuary this is the peak spring-tidal discharge, which (according to Pethick (1984)) also corresponds to bankfull flow. The latter is based on the experience that Q_p just overtops the banks. Another important difference lies in the fact that (unlike in a river) in an estuary the water level is governed by the backwater effect of the ocean, whereas in a river the water level fully depends on the discharge from upstream.

In general, the bank slope of an estuary is very small due to the fine grain sizes and the relatively high shear stress exercised by alternating tidal flows with a peak velocity in the order of 1.0 m/s. If erosion occurs near the toe of the bank, then the erosion will propagate sideways to maintain a stable slope. This widening process often takes place through bank failure. Since the slope is flat the widening is several times more than the deepening. The widening, however, reduces the flow velocity and thus the sediment transport capacity of the stream, leading to a new equilibrium. Hence, an increase in transport capacity of the stream eventually leads to a new equilibrium with a wider channel.

The widening through bank failure is rapid. The opposite process is slower. The building up of a bank by sedimentation, starting from the toe of the banks, may take months if not years. Hence the dynamic equilibrium that is reached as a balance between erosion and sedimentation lies nearer to the maximum eroded profile than to the minimum (silted up) profile. Thus the width of a stream mainly reflects the situation of its maximum eroding capacity. In a river the width is determined by the bankfull flow, in an estuary by the peak spring-tidal discharge.

Observations in excavated tidal canals in Indonesia (at Karang Agung, in the Banyuasin estuary, South Sumatra where the author carried out field surveys in 1989) illustrate this process. An initially prismatic (constant cross section) excavated dead-end canal is seriously eroded at the mouth as high tidal flows enter and leave the canal. The mouth grows deeper after which the banks collapse. At the upstream dead end of the canal where tidal velocities are almost non-existent, sedimentation occurs. The canal thus gradually acquires a funnel shape.

In the mouth of an estuary two different media interact: the ocean, in which the movement of water and sediment has a three-dimensional character and the estuary where the motion is primarily one-dimensional (see Ippen and Harleman, 1966). This interaction often leads to the formation of a shallow area or bar. This shallow reach urges the estuary to become wide and influences the depth of the upstream reaches of the estuary.

It is beyond the scope of this book to go into details regarding the morphological processes that determine channel development. D'Alpaos et al. (2005) have done pioneering work in this field. In this book, we make use of the particular exponential shape of alluvial estuaries that is further discussed in Section 2.2. Equations describing the exponential variation of width and cross-sectional area are essential to the approach followed in this book, and serve the purpose of the 'seventh equation.' But before we do that, we shall see how the St. Venant's equations can be written as functions of tidal velocity and depth.

2.1.3 The one-dimensional equations for depth and velocity

In the following sections, the only dependent variables used in the equations for conservation of momentum and mass of water are U, Z_b, h, and B. The main state variables of interest are the flow parameters: the velocity (U) and the water level ($Z = Z_b + h$). To obtain the St. Venant equations expressed in these variables, the variables A and Q have to be eliminated from Equations 2.1 and 2.2.

Equation 2.1 is the one-dimensional equation of conservation of momentum for water integrated over the cross section. To eliminate the discharge Q and the cross-sectional area A from the equation, use is made of Equations 2.2, 2.7, and 2.8. The following steps are taken

$$\frac{\partial Q}{\partial t} = A\frac{\partial U}{\partial t} + U\frac{\partial A}{\partial t} = A\frac{\partial U}{\partial t} - \frac{U}{r_S}\frac{\partial Q}{\partial x} \tag{2.17}$$

$$\alpha_S\frac{\partial(Q^2/A)}{\partial x} = \alpha_S\left(2U\frac{\partial Q}{\partial x} - U^2\left(h\frac{\partial B}{\partial x} + B\frac{\partial h}{\partial x}\right)\right) \tag{2.18}$$

Intermezzo 2.1:

Substitution of Equations 2.17 and 2.18 in Equation 2.1 yields:

$$A\frac{\partial U}{\partial t} + \left(2\alpha_S - \frac{1}{r_S}\right)U\frac{\partial Q}{\partial x} - \alpha_S U^2\left(h\frac{\partial B}{\partial x} + B\frac{\partial h}{\partial x}\right)$$
$$+ gA\frac{\partial(h + Z_b)}{\partial x} + gA\frac{h}{2\rho}\frac{\partial \rho}{\partial x} + gA\frac{U|U|}{C^2 h} = 0$$

Since $Q = Q(U,B,h)$, elaboration of $\partial Q/\partial x$ yields:

$$A\frac{\partial U}{\partial t} + \left(2\alpha_S - \frac{1}{r_S}\right)U\left(A\frac{\partial U}{\partial x} + Uh\frac{\partial B}{\partial x} + UB\frac{\partial h}{\partial x}\right)$$
$$- \alpha_S U^2\left(h\frac{\partial B}{\partial x} + B\frac{\partial h}{\partial x}\right) + gA\frac{\partial(h + Z_b)}{\partial x} + gAI_r + gA\frac{U|U|}{C^2 h} = 0$$

where I_r is the water level residual slope resulting from the density gradient. Rearrangement yields:

$$\frac{\partial U}{\partial t}+\left(2\alpha_S-\frac{1}{r_S}\right)U\frac{\partial U}{\partial x}+\left(\alpha_S-\frac{1}{r_S}\right)\frac{U^2}{h}\frac{\partial h}{\partial x}+\left(\alpha_S-\frac{1}{r_S}\right)\frac{U^2}{B}\frac{\partial B}{\partial x}$$
$$+g\frac{\partial(h+Z_b)}{\partial x}+gI_r+g\frac{U|U|}{C^2\,h}=0$$

After introduction of the Froude number, $\boldsymbol{F}=U/\sqrt{(gh)}$, this equation can be modified into:

$$\frac{\partial U}{\partial t}+\left(2\alpha_S-\frac{1}{r_S}\right)U\frac{\partial U}{\partial x}+g\left(\boldsymbol{F}^2\left(\alpha_S-\frac{1}{r_S}\right)+1\right)\frac{\partial h}{\partial x}$$
$$+g\frac{h}{B}\boldsymbol{F}^2\left(\alpha_S-\frac{1}{r_S}\right)\frac{\partial B}{\partial x}+g\frac{\partial Z_b}{\partial x}+gI_r+g\frac{U|U|}{C^2\,h}=0$$

Substitution of Equations 2.17 and 2.18 in Equation 2.1 leads to (for details see Intermezzo 2.1):

$$\frac{\partial U}{\partial t}+\left(2\alpha_S-\frac{1}{r_S}\right)U\frac{\partial U}{\partial x}+g\left(\boldsymbol{F}^2\left(\alpha_S-\frac{1}{r_S}\right)+1\right)\frac{\partial h}{\partial x}$$
$$+g\frac{h}{B}\boldsymbol{F}^2\left(\alpha_S-\frac{1}{r_S}\right)\frac{\partial B}{\partial x}+g(I_b-I_r)+g\frac{U|U|}{C^2h}=0 \qquad (2.19)$$

where $\boldsymbol{F}$ is the Froude number: $\boldsymbol{F}=U/\sqrt{(gh)}=U/c_0$, c_0 being the celerity of propagation of a progressive wave (see Imberger's book on Environmental Fluid Dynamics), I_b is the bottom slope, and I_r is the residual slope due to the density gradient. The Froude number in alluvial streams is smaller than unity and generally much smaller: in the order of 0.1. Knowing that $\alpha_S\approx 1$, and that $\boldsymbol{F}^2\ll 1$, the terms containing $\boldsymbol{F}^2(\alpha_S-1)$ may be disregarded. Hence, Equation 2.19 may be simplified into:

$$\frac{\partial U}{\partial t}+\left(2\alpha_S-\frac{1}{r_S}\right)U\frac{\partial U}{\partial x}+g\frac{\partial h}{\partial x}+g(I_b-I_r)+g\frac{U|U|}{C^2h}=0 \qquad (2.20)$$

Scaling the equation

To assess the order of magnitude of the terms in Equation 2.20 let us define a set of dimensionless numbers:

$$U^* = \frac{U}{\upsilon}$$

$$h^* = \frac{h}{\bar{h}}$$

$$x^* = \frac{x}{\lambda}$$

$$t^* = \frac{t}{T}$$

where υ is the amplitude of the tidal velocity, T is the tidal period, λ is the length of the tidal wave (note that $\lambda = cT$) and $\bar{h}$ is the average depth. Equation 2.20 then becomes:

$$\frac{\upsilon}{T}\frac{\partial U^*}{\partial t^*} + \left(2\alpha_S - \frac{1}{r_S}\right)\frac{\upsilon^2}{\lambda}U^*\frac{\partial U^*}{\partial x^*} + g\frac{\bar{h}}{\lambda}\frac{\partial h^*}{\partial x^*} + g(I_b - I_r) + g\frac{\upsilon^2}{\bar{h}}\frac{U^*|U^*|}{C^2 h^*} = 0 \quad (2.21)$$

and with $\lambda = cT$, $c \approx \sqrt{(g\bar{h})}$, and $\boldsymbol{F} = \upsilon/c$:

$$\frac{\partial U^*}{\partial t^*} + \boldsymbol{F}U^*\frac{\partial U^*}{\partial x^*} + 2\left(\alpha_S - \frac{r_S + 1}{2r_S}\right)\boldsymbol{F}\,U^*\frac{\partial U^*}{\partial x^*} + \frac{1}{\boldsymbol{F}}\frac{\partial h^*}{\partial x^*} + \frac{gT}{\upsilon}(I_b - I_r) + \frac{gT\upsilon}{C^2\bar{h}}\frac{U^*|U^*|}{h^*} = 0 \quad (2.22)$$

All scaled variables in Equation 2.22 now have an order of magnitude of 1 and the relative importance of the terms are determined by their dimensionless coefficients. Thus we can see from Equation 2.22 that, since $\boldsymbol{F} < 1$, the second term (the advection term) is small as compared to the fourth term (the depth gradient term). As a result, the advection term is often entirely neglected. We are not neglecting the term here however. Although on an average the second term is small, this may not be true at certain moments during the tidal cycle when the dimensionless velocity gradient can be larger than unity. Hence we retain the term, unless there are specific reasons not to do so. What we can do, however, is neglect the third term containing the effect of α_S and r_S on the advection term. Since in alluvial estuaries both α_S and r_S are close to unity, the third term is an order of magnitude smaller than the second term and, as a result, this term may be disregarded. Equation 2.19 thus becomes:

$$\frac{\partial U}{\partial t} + U\frac{\partial U}{\partial x} + g\frac{\partial h}{\partial x} + g(I_b - I_r) + g\frac{U|U|}{C^2 h} = 0 \quad (2.23)$$

The conservation of mass equation for water, Equation 2.2, is dealt with in a similar way by making use of Equations 2.7 and 2.8:

$$r_S\left(h\frac{\partial B}{\partial t}+B\frac{\partial h}{\partial t}\right)+UB\frac{\partial h}{\partial x}+hB\frac{\partial U}{\partial x}+hU\frac{\partial B}{\partial x}=0 \tag{2.24}$$

In geometric terms, at a fixed location, the width B is a sole function of h. Hence the first term can be written as:

$$h\frac{\partial B}{\partial t}=h\frac{\mathrm{d}B}{\mathrm{d}h}\frac{\partial h}{\partial t} \tag{2.25}$$

If we assume a trapezoidal cross-sectional shape with a side slope i, then Equation 2.24 can be written as:

$$r_S\left(1+\frac{2h}{iB}\right)\frac{\partial h}{\partial t}+U\frac{\partial h}{\partial x}+h\frac{\partial U}{\partial x}+\frac{hU}{B}\frac{\partial B}{\partial x}=0 \tag{2.26}$$

In estuaries, the variation of the cross-sectional area in time is mainly caused by the variation of the water level and much less by the variation in width ($B \gg h$). Consequently, the term h/iB is normally small with respect to unity. Hence in the first term of Equation 2.26 the effect of side slope is disregarded, or considered part of the storage width ratio.

$$r_S\frac{\partial h}{\partial t}+U\frac{\partial h}{\partial x}+h\frac{\partial U}{\partial x}+\frac{hU}{B}\frac{\partial B}{\partial x}=0 \tag{2.27}$$

Assuming that values for C and r_S can be determined independently through measurements, Equations 2.23 and 2.27 form a set of two equations with four unknowns: U, h, I_b, and B from which A and Q have been eliminated. Therefore, if we want to solve these equations, we still require two geometric relations to determine the width and the bottom slope. This is done in Section 2.2.

2.1.4 The effect of density differences and tide

Until this point, the derivations made could apply to any channel of varying shape, whether it is a river, a canal, a lagoon, or an estuary, as long as it can be described as a one-dimensional system. In this section, the typical hydraulic characteristics are presented of a tidal estuary with salt intrusion of the well-mixed type. These characteristics are twofold:

- the effect of density differences
- the tidal movement

The density effect

In the downstream part of an estuary, in the period of the year when the upstream fresh discharge is small, the tidal flows substantially dominate the fresh water flow with the consequence that the water in that part of the estuary turns saline. If the fresh flow from upstream is small, the mixing in the estuary is generally good and the salinity decreases gradually from the estuary mouth in an upstream direction. In the above equations, the residual slope I_r has been used to account for the effect of the density gradient $\partial\rho/\partial x$ on the momentum balance; see Equation 2.23. In the Intermezzo 2.2, the expression for this density term is derived from the water pressure force, resulting in:

Intermezzo 2.2:

The force per unit mass of water F exercised by the water pressure is defined as:

$$F(x,z) = -\frac{1}{\rho}\frac{\partial\left(\rho g(h - Z_b - z)\right)}{\partial x}$$

where z is the vertical ordinate. The force can be split up into three components (Van Os and Abraham, 1990):

$$F(x,z) = -g\frac{\partial\left(h + Z_b\right)}{\partial x} - \frac{gh}{2\rho}\frac{\partial\rho}{\partial x} - \frac{g}{\rho}\left(\frac{h}{2} + Z_b - z\right)\frac{\partial\rho}{\partial x}$$

The separation in three terms has been done in a way that only the third term is z-dependent. At the water surface, where $z = Z_b + h$, the third term is equally large as the second term, but of the opposite sign; at the bottom, the third term equals the second term. The second term is independent of z, because in a well-mixed estuary it is assumed that $\partial\rho/\partial x$ is not z-dependent. Integration over the depth from Z_b to $Z_b + h$ and division by the depth h yields the depth average water pressure force per unit mass:

$$F(x) = -g\frac{\partial\left(h + Z_b\right)}{\partial x} - \frac{gh}{2\rho}\frac{\partial\rho}{\partial x} - \frac{g}{\rho h}\frac{\partial\rho}{\partial x}\int_{Z_b}^{Z_b+h}\left(h/2 + Z_b - z\right)dz$$

The first term represents the water slope. The second term is a density driven force which points upstream (in a positive estuary). The last term equals zero, as—in this term—the pressure varies linearly from a downstream directed pressure at the surface to an equal but opposed upstream directed pressure at the bottom. Hence the second term is the resulting upstream pressure as a result of the density gradient. This does not mean that the third term is unimportant. In some cases it has a dominant effect on salinity intrusion through gravitational circulation. As a result of a net upstream water pressure near the bottom, and a net downstream water pressure near the surface (see Figure 2.2), there exists a time-average upstream flow near the bottom and a net downstream flow near the surface.

$$\frac{\partial U}{\partial t} + U\frac{\partial U}{\partial x} + g\frac{\partial h}{\partial x} + \frac{gh}{2\rho}\frac{\partial \rho}{\partial x} + g\frac{\partial Z_b}{\partial x} + g\frac{U|U|}{C^2 h} = 0 \tag{2.28}$$

Compared to the other terms in the cross-sectional average conservation of momentum equation, the density term is small. Scaling of the terms in Equation 2.28 leads to the conclusion that the ratio of the third and fourth term is of the order $h\Delta\rho/(H\rho)$, where H is the tidal range and $\Delta\rho$ is the density difference of ocean and river water. In open sea estuaries, this ratio is about $(1025\text{–}1000)/2000\, h/H = 0.0125\, h/H$. Even though H/h is supposed to be less than unity, this is still a small number. However, the third term (the water slope) alternates (with the tide) between a positive and a negative value, whereas the fourth term always exercises a pressure in an upstream direction (in a normal, 'positive,' estuary). This pressure is counteracted by a residual water level slope amounting to 1.25 percent of the estuary depth over the salt intrusion length L, the distance from the mouth to the point where the estuary water is fresh. If, for example, the estuary depth is 8 m then the water level rise amounts to 0.1 m over the salt intrusion length L. In formula this yields the following expression for the residual slope I_r:

$$I_r \approx \frac{\Delta\rho}{2\rho}\frac{\bar{h}}{L} = 0.0125\frac{\bar{h}}{L} \tag{2.29}$$

The same relation is obtained by equating the shadowed areas in Figure 2.2: $\rho h\Delta h = \Delta\rho h^2/2$. The difference in water level is required to balance the hydrostatic

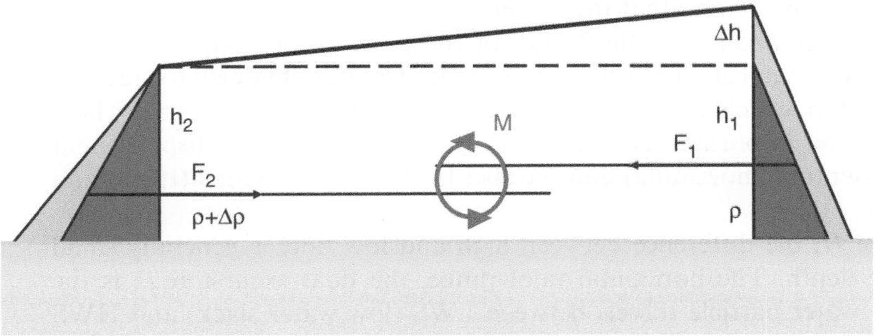

Figure 2.2 Resultant hydrostatic forces driving vertical net circulation.

forces. The two forces F_1 and F_2 that make equilibrium in the horizontal plane per unit width over the salt intrusion length L are:

$$F_1 = \frac{1}{2}\rho_1 g h_1^2 \tag{2.30}$$

and

$$F_2 = \frac{1}{2}\rho_2 g h_2^2 \tag{2.31}$$

where the subscripts 1 and 2 indicate the upstream and downstream ends of the salt intrusion length L. Since $(\rho_2 = \rho + \Delta\rho) > (\rho_1 = \rho)$, there can only be equilibrium if $h_1 > h_2$. However, the two forces, although equal and opposite, exert a momentum that drives the gravitational circulation described in Intermezzo 2.2. Since the arm of the momentum is $\Delta h/3$, the moment M, exercised per unit volume of water per unit width (Lh), equals:

$$M = \frac{\frac{1}{3}\frac{h}{2\rho}\frac{\partial\rho}{\partial x}L\frac{1}{2}\rho g h^2}{Lh} = \frac{1}{12}\frac{\partial\rho}{\partial x}gh^2 \tag{2.32}$$

The moment drives the vertical mixing process, called gravitational circulation, which is further discussed in Section 4.2.

Tidal characteristics

A second characteristic of the lower part of an estuary is that both the water level fluctuations and the velocity of the water are tidally dominated and that U and h vary according to periodic functions. In the mouth of an estuary the water level rises and falls periodically. This cyclic rise and fall produces a tidal wave of a primarily one-dimensional character which travels up the estuary. The period T of a tidal wave is generally so long that the wavelength $\lambda = Tc$ (c is the celerity of the wave) is usually much larger than the length of the estuary considered. To show that exceptions prove the rule, the tidal influence in The Gambia estuary reaches a distance of 500 km which equals approximately the tidal wave length. These long waves have the important characteristic that the associated displacement of the water is essentially horizontal and parallel to the estuary banks (Ippen and Harleman, 1966).

The tidal range H, the difference between high and low tide, is generally small compared to the depth. The horizontal tidal range, the tidal excursion E, is the distance which a water particle travels between LWS (low water slack) and HWS (high water slack). Generally, the tidal excursion is substantially larger than the estuary width and small in relation to the estuary length. If this is not the case, then the estuary is so wide that it loses its one-dimensional character and should rather be considered as a lagoon, a bay, or a part of the estuary mouth. This brings us to the following inequalities:

$$H < h \ll B < E \ll \lambda \tag{2.33}$$

The volume of water entering the estuary between LWS and HWS is known as the flood volume P_t, which in the literature is often referred to as the tidal prism:

$$P_t = \int_{\text{LWS}}^{\text{HWS}} Q(0,t)\mathrm{d}t \approx A_0 E_0 \tag{2.34}$$

The product of the tidal excursion E_0 and the cross-sectional area A_0 at the estuary mouth appears to be a good approximation for the tidal prism (see Section 2.3). For the analysis of mixing of fresh water and saline water, the ratio between the amount of fresh water and salt water entering the estuary is important. This ratio, in the Dutch literature referred to as Canter Cremers' estuary number N, is defined as:

$$N = \frac{Q_f T}{P_t} \tag{2.35}$$

where Q_f is the fresh river discharge which enters the estuary during the tidal period T.

A more significant estuary number is the Estuarine Richardson number N_R (Fischer et al., 1979) which represents the balance between, on the one hand, the potential energy per tidal period needed for mixing against buoyancy (or the potential energy gained by making fresh water saline): $E_m = \Delta\rho Q_f T g(h/2)$, and, on the other hand, the kinetic energy per tidal period supplied by the tidal current for realizing the mixing $E_t = 0.5\rho A_0 E_0 \upsilon_0^2$, where υ_0 is the amplitude of the tidal flow velocity at the estuary mouth:

$$N_R = \frac{E_m}{E_t} = \frac{\Delta\rho}{\rho}\frac{ghQ_f T}{A_0 E_0 \upsilon_0^2} = \frac{\Delta\rho}{\rho}\frac{ghQ_f T}{P_t \upsilon_0^2} \tag{2.36}$$

Hence $N_R = N/\boldsymbol{F}_\mathbf{d}$, where $\boldsymbol{F}_\mathbf{d}$ is the densimetric Froude number defined as $\boldsymbol{F}_\mathbf{d} = (\rho/\Delta\rho)\upsilon_0^2/(gh)$. Fischer et al. (1979) states: 'If N_R is very large, we expect the estuary to be strongly stratified and the flow to be dominated by density currents. If N_R is very small, we expect the estuary to be well mixed, and we might be able to neglect density effects. Observations of real estuaries suggest that, very approximately, the transition from a well mixed to a strongly stratified estuary occurs in the range $0.08 < N_R < 0.8$.'

Harleman and Thatcher (1974) used a similar estuary number, which is the reciprocal value of N_R. Prandle (1985) has a number similar to Harleman and Thatcher, which he also based on energy considerations, but based on the ratio of the energy dissipated by friction over the salt intrusion length $E_d = 4/(3\pi)$ $(g/C^2)\rho \upsilon^3 LBT$ to the potential energy E_m gained by mixing. This yields Prandle's estuary number N_P:

$$N_P = \frac{8}{3}\frac{g}{C^2}\frac{L}{h}\frac{1}{N_R} \qquad (2.37)$$

In addition to the Estuarine Richardson number, it accounts for friction and the salt intrusion length to depth ratio. Particularly the inclusion of the salt intrusion length makes this number a strong indicator for stratification. Both a large L/h ratio and a small value of N_R correspond with a well-mixed estuary; so a large value of N_P corresponds with a well-mixed estuary and a small value implies stratification. Because both the friction and the intrusion length are difficult to determine *a priori*, this is not a very useful estuary number to predict stratification, however.

The tidal wave

Three types of tidal waves can be distinguished:

1. a standing wave
2. a progressive wave
3. a wave of mixed type

Only the latter type of wave occurs in an alluvial estuary, which gradually tapers into an alluvial river. A purely standing wave requires a semi-enclosed body where the tidal wave is fully reflected. Since an alluvial estuary gradually changes into a river, this does not apply. Standing waves can only occur in non-alluvial estuaries or in estuaries where a closing structure has been constructed and then only in the vicinity of the structure since the reflected wave, moving in downstream direction, quickly loses energy due to friction and widening of the channel (see also Jay, 1991). A standing wave reaches extreme water levels simultaneously along the estuary. Consequently, the 'apparent' celerity c tends to infinity (as extreme water levels occur everywhere at the same time, it appears as if the celerity is infinitely large). In addition, HWS coincides with HW (high water) and LWS coincides with LW (low water). The phase lag φ between the fluctuation of the water level Z and the flow velocity U is $\pi/2$ (see Figure 2.3). In Figure 2.3, the positive direction of flow is in the upstream direction.

A purely progressive wave only occurs in a frictionless channel of constant cross section and infinite length. Estuaries do not belong to that category. In the event of a progressive wave, water level and stream velocity are in phase (i.e. high water occurs at the same time as the maximum flow velocity). The phase lag φ

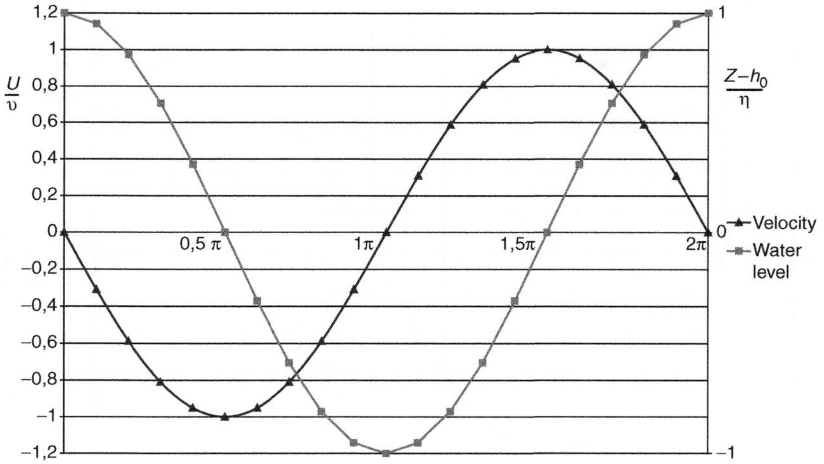

Figure 2.3 A standing wave.

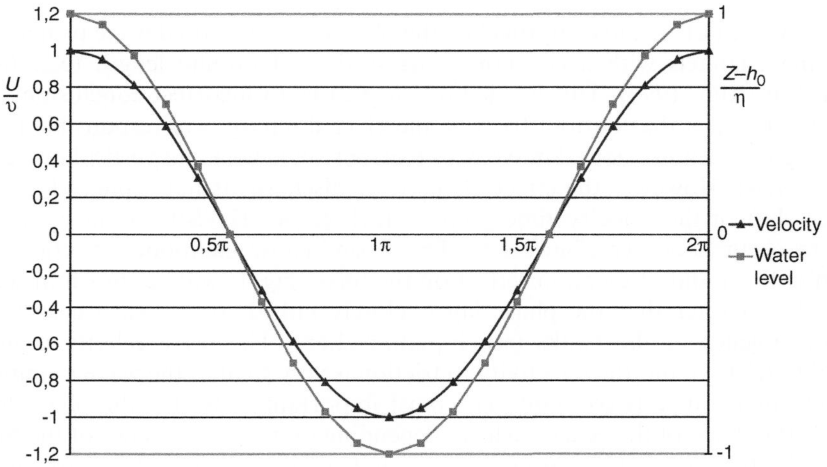

Figure 2.4 A progressive wave.

between water level and flow velocity is zero and the wave celerity $c = c_0\ \sqrt{(gh)}$ (see Figure 2.4).

None of these extreme situations occur in an alluvial estuary. The tidal wave in an estuary is of a 'mixed' type, with a phase lag φ between 0 and $\pi/2$ (see Figure 2.5). This means that in an alluvial estuary HWS occurs after HW and before mean tidal level; and LWS occurs after LW and before mean tidal level.

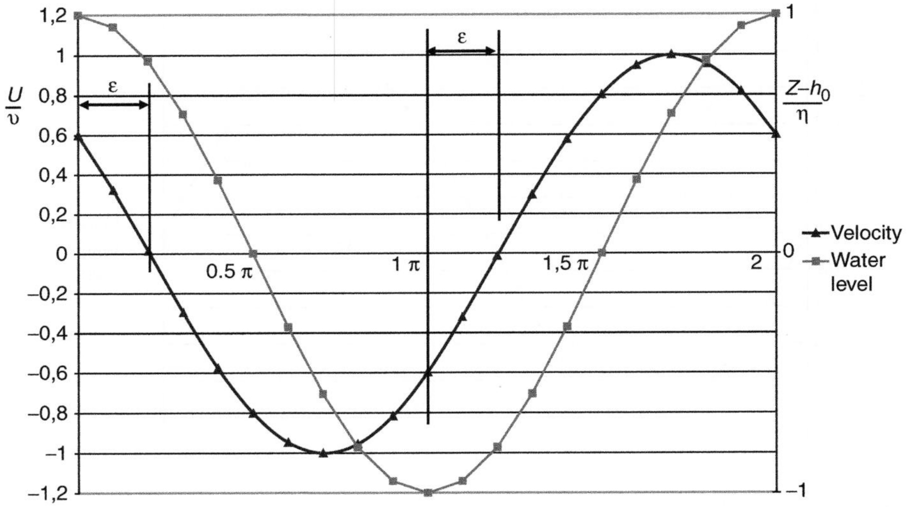

Figure 2.5 A wave of the mixed type showing the phase lag between HW and HWS, and between LW and LWS.

Several researchers have approached the phenomenon of tidal wave propagation analyzing channels with a constant cross section of infinite length (e.g. Ippen, 1966a; Van Rijn, 1990). This has led in some cases to incorrect conclusions. Van Rijn (1990) states that bottom friction and river discharge are responsible for the phase lag between horizontal movement (current velocities) and vertical movement (water levels). However, the effect of the river discharge is not a phase lag, but a vertical shift of the velocity–time graph, which causes HWS to occur earlier and LWS to occur later (see Figure 1.5). The second cause mentioned (friction) only has an indirect and often minor effect on the phase lag. It will be shown in further on in this chapter that the phase lag is closely linked to the wave celerity and the convergence of the banks (see Equation 2.88). The wave celerity is indeed related to friction, but the sensitivity to friction is very small if the wave is amplified or if the tidal range is constant. The most important cause for the phase lag to occur is the shape of the estuary which, depending on the convergence of the banks, causes the tidal wave to gain energy per unit width as it travels upstream. The relationship for the phase lag as a function of bank convergence and wave celerity is derived in Section 2.3.

Here, we introduce the phase lag ε between HW and HWS (see Figure 2.5), or between LW and LWS, which is related to φ as: $\varepsilon + \varphi = \pi/2$. The phase lag ε, although disregarded by many authors, is a very important parameter in tidal hydraulics, characterizing the hydraulics of an estuary. It is closely related to estuary shape. The phase lag is primarily a function of the ratio between bank convergence and tidal wave length (see Section 2.3). In alluvial estuaries this phase lag is always between zero and $\pi/2$, but typically in the order of 0.3, resulting in a time lag between HW and HWS of around 30–45 min for a semi-diurnal tide. We also define the dimensionless Wave-type number: $N_E = \sin(\varepsilon)$ which defines the wave type in an estuary. The Wave-type number is always between zero and unity. If it is close to unity the wave is a progressive wave and the estuary is a prismatic channel. If it is close to zero the wave is a standing wave and the estuary looks like a tidal embayment. Estuary shape being so important in tidal hydraulics, the next section will elaborate the topography of alluvial estuaries.

2.2 THE SHAPE OF ALLUVIAL ESTUARIES

2.2.1 Classification on estuary shape

Many authors have used prismatic (constant width) channels to study tidal hydraulics and salt intrusion in estuaries. Here, we use a more general approach where we allow the cross section and the width to vary along the estuary axis according to exponential functions. The rate of longitudinal convergence is determined by a length scale called the convergence length. A channel with constant cross section is a special type of estuary with an infinitely long convergence length.

It was mostly for reasons of mathematical convenience that researchers used prismatic channels, but there were also practical reasons. Many tests were done on the basis of laboratory experiments and laboratory flumes are generally prismatic. Moreover, several real-life problems that early researchers had to analyze concerned man-made shipping access channels such as the Rotterdam Waterway, which has a constant cross section. A vast amount of literature on salt water intrusion deals with prismatic channels. Until 1992, virtually all formulae that existed to determine the salt intrusion length had been derived for prismatic channels. As will be shown in Chapter 5 these equations perform very poorly in natural estuaries.

Few estuaries can be adequately described as having a prismatic channel. Therefore, throughout this book, use is made of a cross-sectional area that varies exponentially with the distance. The justification of this assumption is presented in Section 2.2.2. Since the positive x-direction is chosen in upstream direction, the formula reads:

$$A = A_0 \exp\left(-\frac{x}{a}\right) \tag{2.38}$$

where A_0 is the cross-sectional area at $x = 0$. The parameter a is defined as the cross-sectional convergence length (a is the distance from the mouth at which the tangent through the point $(0, A_0)$ intersects the x-axis). Similarly the assumption that the width varies exponentially yields the equation:

$$B = B_0 \exp\left(-\frac{x}{b}\right) \tag{2.39}$$

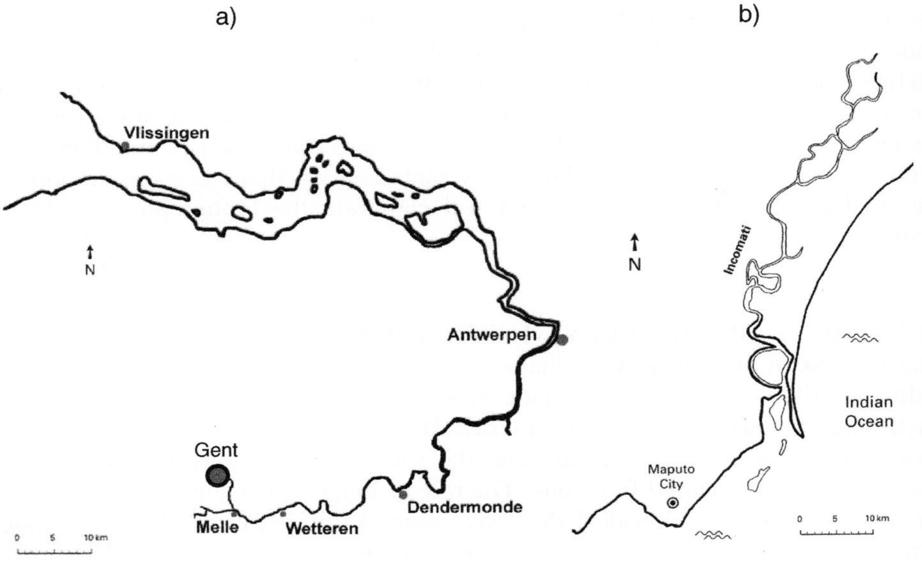

Figure 2.6 Sketch of the Schelde (a) and Incomati (b) estuaries.

where the coefficient b is the width convergence length. Combination of Equation 2.38 with 2.39 leads to an expression for the depth:

$$h = h_0 \exp\left(\frac{x(a-b)}{ab}\right) \tag{2.40}$$

It follows from Equation 2.40 that, if a is larger than b, the depth increases exponentially; if a is less than b, the depth decreases exponentially. If a and b differ substantially an unrealistic situation is reached. As we shall see further on, in real estuaries the convergence lengths do not differ much. In the special case where the two convergence lengths are equal, $a = b$, the depth is constant along the estuary: $h = h_0$. Figure 2.6 shows a sketch of two estuaries that we shall use for illustration purposes in the text: The Schelde in Belgium and The Netherlands, and the Incomati in Mozambique. Figures 2.7 and 2.8 show the geometry of these estuaries of which measurements of A, B, and h are available, plotted on a semi-log scale. The individual marks of the depth are based on scattered point observations where soundings were made during salt measurements. They do not reflect the average depth. The observations of the cross-sectional area are the result of a complete echo-sounding. It can be clearly seen that the trends in cross-sectional area and the width conform very neatly with Equations 2.38 and 2.39. The Incomati has an inflection in its shape 14 km from its mouth, while the Schelde becomes more shallow at 110 km where several tributaries branch off. In spite of these

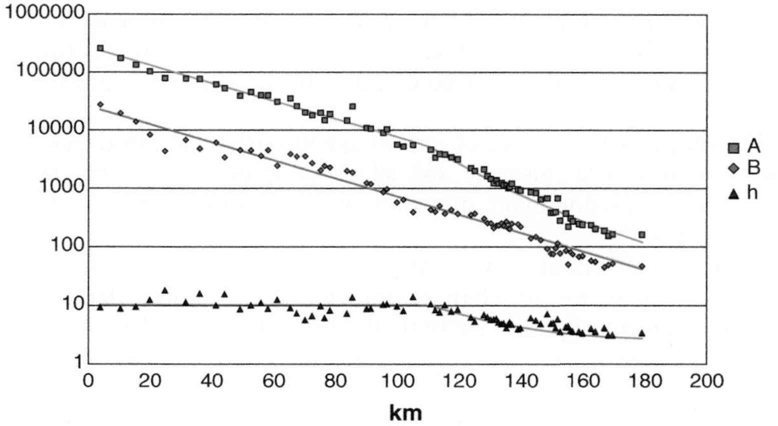

Figure 2.7 Semi-logarithmic plot of the geometry of the Schelde estuary: A is the cross-sectional area in m^2, B is the width in m, h is the cross-sectional average depth in m.

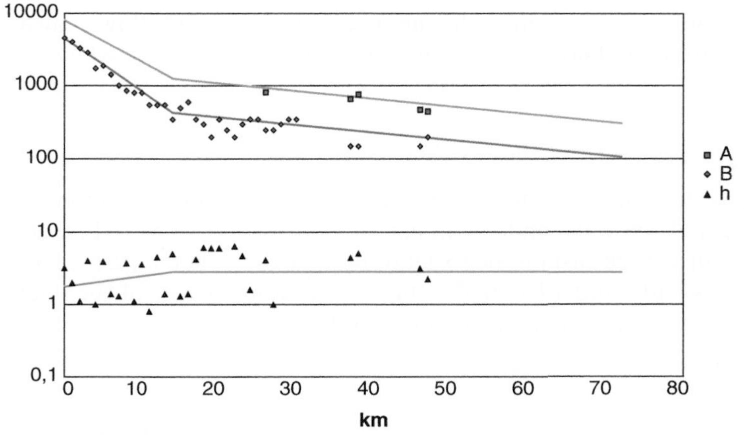

Figure 2.8 Semi-logarithmic plot of the geometry of the Incomati estuary: A is the cross-sectional area in m^2, B is the width in m, h is the cross-sectional average depth in m.

irregularities, the geometry can be described very well by Equations 2.38–2.40. Also it appears that a and b are almost, or exactly equal. The question is: what are the factors determining estuary shape?

The shape of estuaries depends on several factors such as:

- tidal movement. Both the vertical and horizontal displacements. The main variable determining the downstream boundary condition for tidal movement is the tidal range H, a good indicator for the strength of the tidal movement.

- river floods. The morphologic activity of the river can strongly influence the estuary shape; if river floods are large then the estuary gets a more riverine character and a more prismatic shape. A good indicator for a river flood is the bankfull discharge Q_b.
- wave action. The shape of the mouth of an estuary in particular can depend strongly on wave action. The existence and shape of spits, bars, or barrier islands depends on the predominant direction of wave attack and on the magnitude of the waves.
- storm action. Storms can change the configuration of the estuary mouth considerably. The permanence of changes inflicted by a storm depends on the amount of sediment supplied by both the littoral zone and the river itself.

Of these factors, the tidal range is the easiest variable to determine. Moreover, Hayes (1975) (who followed the classification proposed by Davies (1964)) stated that 'the tidal range had the broadest effect in determining large-scale differences in morphology of sand accumulation' and that a classification of estuaries could best be based on the tidal range.

Pethick (1984), although recognizing the importance of upland flows, also followed the classification of Davies (1964) based on the tidal range, which is summarized as:

Micro-tidal estuaries

When the tidal range is less than 2 m, the estuarine processes are dominated by both the upland discharge and the wave and storm action from the sea. The sediments carried by the upland discharge sustain the formation of a delta, whereas the waves produce spits, barrier islands, and a bar-built estuary. The convergence of the tidal channel is small (the convergence lengths a and b are long).

Meso-tidal estuaries

Estuaries with a tidal range between 2 and 4 m experience such strong tidal action that a marine delta can no longer be shaped. Instead, two shallow reaches are formed both upstream and downstream of the estuary mouth which are called the flood-tide and the ebb-tide delta, respectively.

Macro-tidal estuaries

In macro-tidal estuaries the tidal range is over 4 m. The tide produces strong tidal currents which may extend for hundreds of kilometres inland. They do not possess ebb-tide or flood-tide deltas but have a pronounced funnel shape with a strong convergence (the convergence lengths are short).

Although this classification has the advantage of simplicity and probably is adequate for a physical geographer (it allows the classification of the mouth of the estuary), it is not sufficiently accurate for the engineer who is interested in the estuarine morphology.

Estuaries, according to Dyer (1973), are sediment traps. The sediment supplied by the river floods is deposited in the estuary as soon as the channel becomes wider and the flow velocities decrease. In addition, the gravitational circulation sketched in Figure 2.2 continuously supplies fine marine sediments which move upstream near the estuary bottom to be eventually deposited at the limit of the salt intrusion. Only the river floods are able to flush out the sediments which have accumulated over the years.

Wright et al. (1975), who studied the morphology of the Ord estuary in Western Australia, a typical macro-tidal estuary, formulated it thus: 'In a channel of uniform cross section, the upstream increase in tidal asymmetry and accompanying flood-dominated bed load transport would, in the absence of significant riverine flow, lead to an upstream accumulation of sediment to clog the channel; only during a river flood would this sediment be flushed.' Hence, a substantial river flow is required to maintain a channel with a long convergence length.

In a prismatic channel, the tidal flow velocities increase in a downstream direction. Consequently, erosion dominates at the downstream end of such an estuary. If the river discharge and its sediment load are small, this erosion is not replenished by riverine sediments. In such a case the estuary expands to form a funnel shape. If, on the other hand, the upland discharge and sediment load are large, then the channel is only stable if the convergence length is long. If the convergence length would be short, the channel would soon fill up with riverine sediments. Consequently, a high upland discharge induces a channel with a long convergence length. A large tidal range however induces a channel with a short convergence length, in agreement with the classification of Pethick (1984). Hence it is the proportion between these two actions which determines estuary shape.

In Table 2.1, a number of characteristics related to estuary shape are presented with a qualification of how they behave in a predominantly funnel shaped or prismatic channel. If the proportion of upland flood discharge to tidal range is small the estuary has a predominantly funnel shape, if the proportion is large the estuary is predominantly prismatic.

Therefore a better classification would be based on the proportion of the upland discharge to the tidal range; or, to make it dimensionless, the ratio of a river flood

Table 2.1 Characteristics of funnel-shaped and prismatic estuaries

Shape	Bay shape	Funnel shape	Prismatic shape
Character	Marine	Estuarine	Riverine
Convergence length (a)	0	Short	∞
HW–HWS phase lag (ε)	0	$0 < \varepsilon < \pi/2$	$\pi/2$
Wave type	Standing	Mixed	Progressive
Salt intrusion	Saline	Well-mixed	Stratified
$Q_b T/P_t$	0	Small	Large

volume entering the estuary during a tidal period to the flood volume P_t, the volume of ocean water entering the estuary during a tidal period. This ratio is a Canter Cremers number for an upland flood discharge. The upland discharge to be used in the classification is the characteristic annual flood discharge. In regime theory, the flood discharge that determines the shape of an alluvial channel is the bankfull discharge Q_b. For the purpose of classification this discharge is a good selection, not because it is better than any other criterion, but because it is an objective criterion and it is relatively easy to determine. Riggs (1974) making use of investigations by Leopold et al. (1964), stated that bankfull stage has a return period of 1.5 years. Personal observations confirm this statement, which may find its explanation in the fact that natural river banks need regular replenishment with bed material for the river to maintain its course.

If a satisfactory relationship can be found between Q_bT/P_t and estuary shape, then that would be a better means for classification than the one proposed by Davies (1964) and Pethick (1984).

2.2.2 Assumptions on the shape of alluvial estuary in coastal plains

The assumptions of an ideal estuary

During the many boat surveys the author carried out during the 1980s in Mozambican and Asian estuaries (Limpopo, Pungué, Maputo, Incomati, Pungué, Lalang, Tha Chin, Chao Phya), it appeared that these estuaries, although quite different in hydrology and geometry, had certain geometric characteristics in common.

In the first place it became clear that, contrary to expectation, the mean depth of the estuaries did not significantly change when moving upstream from the estuary mouth. Although the depth sometimes fluctuated strongly from place to place (deep in bends and shallow in a straight stretch) there did not appear to be an upward or downward bottom slope. This indicated that the depth of flow h was more or less constant with distance.

A second phenomenon observed was that the amplitude of the tidal flow velocity (i.e. the maximum flow velocity) near the mouth was of the same order of magnitude as the maximum flow velocity observed near the limit of the salt intrusion 50–100 km upstream. Even more remarkably, this velocity did not differ much from estuary to estuary; regardless of whether the tidal range was large (such as in the Pungué) or small (as in the Limpopo). In both cases the maximum flow velocity at spring tide was in the order of 1 m/s. The absence of a gradient in the velocity amplitude implies a constant tidal excursion E along the estuary axis. The fact that the peak velocity is similar in different estuaries is the result of similar physical characteristics of alluvial estuaries (discussed earlier in Section 2.1.1) having an almost direct proportionality between Q and A (e.g. Bruun and Gerritsen, 1960).

In addition, although most estuaries experience some degree of tidal damping or amplification, it appeared that the gradient of the tidal range H was modest or, in other words, that the tidal range remained fairly constant along these estuaries, at least in the tidal dominated part of the estuary.

Now it should be observed that these are all coastal plain estuaries unaffected by steep topography. In estuaries where the coastal plain is short with a steep underlying topography another type of estuary occurs, which we shall describe in Section 2.2.3.

The experiences gained during these surveys formed the inspiration to develop a salt intrusion model (Savenije, 1986) based on the following assumptions regarding the shape and hydraulics of an alluvial estuary:

$$h(x) = h_0 \tag{2.41}$$

$$B(x) = B_0 \exp\left(-\frac{x}{b}\right) \tag{2.42}$$

$$H(x) = H_0 \tag{2.43}$$

$$E(x) = E_0 \tag{2.44}$$

The tidal range H and the tidal excursion E are here presented as mere functions of x. This is the situation for a specific tidal wave on a certain day. From day to day the tidal range and the tidal excursion vary with time, and in that sense H and E are functions of time. For a certain tidal wave that travels up an estuary, however, they are merely functions of space. The subscript 0 for the tidal range, the tidal excursion, and the depth indicate that the parameters are constant along the estuary.

Equations 2.41 and 2.42 agree with Equations 2.38–2.40 if $a = b$. These equations correspond to an 'ideal estuary' as described theoretically by Pillsbury (1939, 1956), which in addition to the above geometric conditions requires a constant Chézy coefficient. Apparently a funnel-shaped estuary, with the width obeying an exponential function, is best suited to preserve a constant tidal range and hence to maintain a constant amount of wave energy per unit volume of water. The contracting width tends to increase the tidal range, whereas the friction tends to reduce the tidal range. Dyer (1973) formulates it thus: 'As an estuary narrows towards the head, the tidal range tends to increase upstream because of the convergence but decrease because of friction.' In an ideal estuary (according to Langbein, 1963), the convergence of the estuary banks is just sufficient to balance the damping of the tidal range due to friction.

A constant amount of energy per unit volume of water implies that energy dissipation by friction is balanced by energy gained through convergence. Since in an exponentially shaped estuary the latter is constant, the amount of energy spent per unit volume of water is also constant. The latter is a condition for morphological stability that is also used to describe river channel networks. Rodriguez-Iturbe and Rinaldo (1997; p. 267) use the criterion of 'equal energy expenditure per unit volume' to describe and simulate natural topographies. An ideal estuary is the coastal version of a self-organized river network, the difference lying mainly in the forcing boundary condition. An estuary is forced by the tidal variation at the downstream boundary while a river is forced by the river discharge at the upstream boundary. In the following, a brief theoretical justification of an ideal estuary is presented.

Theoretical justification

The Equations 2.41–2.44 for an ideal estuary can be obtained from the general St. Venant equations as formulated in Equations 2.23 and 2.27, under the following assumptions.

1. Since the Froude number is small, the non-linear convergence term of Equation 2.23 is much smaller than $g\partial h/\partial x$, and hence may be disregarded.
2. The resistance term $gU|U|/(C^2h)$ of Equation 2.23 may be linearized.
3. The velocity of the fresh water discharge U_f is negligible when compared to the tidal velocity amplitude υ.
4. In the downstream part of the estuary, the mean tidal water level Z_0 is independent of x (implying that the residual slope I_r in Equation 2.23 can be disregarded as well).
5. The storage width ratio in Equation 2.27 is close to unity.
6. The water movement (both velocity and water level) can be described by a combination of harmonics.
7. The damping of both the tidal range and the amplitude of the tidal velocity is small. Hence the variation of $H(x)$ and $E(x)$ (proportional to the amplitude of the tidal velocity) with x is small or negligible.

The first six assumptions are not very restrictive and are in fact often made in alluvial estuaries (although we shall use less restrictive assumptions in Section 2.3 and Chapter 3). The seventh assumption may not be made in estuaries that are forced by the topography to be short (See Section 2.2.3). In those estuaries $a \neq b$. In coastal plain estuaries however, particularly in the downstream marine-dominated area, the tidal range and the velocity amplitude are not significantly damped or amplified.

Assumptions 1 and 2 imply that use can be made of the linearized St. Venant equations:

$$\frac{\partial U}{\partial t} + g\frac{\partial Z}{\partial x} + R_L U = 0 \tag{2.45}$$

$$B\frac{\partial Z}{\partial t} + Bh\frac{\partial U}{\partial x} + U\frac{\partial Bh}{\partial x} = 0 \tag{2.46}$$

Equation 2.45 follows from Equation 2.23, where $Z = h + Z_b$ is the water level and R_L is Lorentz's[7] linearized friction factor:

$$R_L = \frac{8}{3\pi} \frac{g}{C^2} \frac{\upsilon}{\bar{h}} \tag{2.47}$$

In Equation 2.46 a substitution has been made of $\partial Z/\partial t = \partial h/\partial t$, since the bottom slope does not vary at the time scale considered.

If we take into account Assumptions 3–7, then the velocity U and the water level Z can be written as sole functions of the Lagrangean variable ξ (Savenije, 1986, after a personal communication by Kranenburg, 1985):

$$U = \upsilon \Phi(\xi - \varepsilon) \tag{2.48}$$

$$Z = \eta \Psi(\xi) + Z_0 \tag{2.49}$$

$$\xi = \omega t - \varphi(x) + \xi_0 \tag{2.50}$$

where η is the tidal amplitude (equal to $H/2$), ε is the phase lag between HW and HWS, ω is the angle velocity ($\omega = 2\pi/\mathrm{T}$), and $\varphi(x)$ is the phase shift resulting from wave propagation. The functions Φ and Ψ are periodic functions with amplitude of unity. Substitution of Equations 2.48–2.50 in Equations 2.45 and 2.46, and some rearrangement yields:

$$\upsilon \frac{\mathrm{d}\Phi}{\mathrm{d}\xi}\frac{\partial \xi}{\partial t} + g\eta \frac{\mathrm{d}\Psi}{\mathrm{d}\xi}\frac{\partial \xi}{\partial x} + R_L \upsilon \Phi = 0 \tag{2.51}$$

$$\upsilon Bh \frac{\mathrm{d}\Phi}{\mathrm{d}\xi}\frac{\partial \xi}{\partial x} + B\eta \frac{\mathrm{d}\Psi}{\mathrm{d}\xi}\frac{\partial \xi}{\partial t} + \frac{\mathrm{d}Bh}{\mathrm{d}x}\upsilon \Phi = 0 \tag{2.52}$$

Further elaboration yields:

$$\omega \left(\upsilon \frac{\mathrm{d}\Phi}{\mathrm{d}\xi} \right) - g \frac{\mathrm{d}\varphi}{\mathrm{d}x} \left(\eta \frac{\mathrm{d}\Psi}{\mathrm{d}\xi} \right) + R_L(\upsilon \Phi) = 0 \tag{2.53}$$

$$-Bh \frac{\mathrm{d}\varphi}{\mathrm{d}x} \left(\upsilon \frac{\mathrm{d}\Phi}{\mathrm{d}\xi} \right) + B\omega \left(\eta \frac{\mathrm{d}\Psi}{\mathrm{d}\xi} \right) + \frac{\mathrm{d}Bh}{\mathrm{d}x}(\upsilon \Phi) = 0 \tag{2.54}$$

[7] The Dutch Nobel prize winner H.A. Lorentz is not normally associated with hydraulic engineering, but he pioneered the numerical approach in hydraulic engineering after his retirement, when he took up the job of predicting the effect of the closure of the Zuiderzee (the main inland sea of The Netherlands) on the tidal variations in the semi-enclosed Waddenzee.

The solution only suffices if Φ and Ψ are solely dependent on ξ and thus if there is a proportionality between the coefficients of the terms of each equation:

$$\omega \propto g\frac{d\varphi}{dx} \propto R_L \tag{2.55}$$

and:

$$Bh\frac{d\varphi}{dx} \propto B\omega \propto \frac{dBh}{dx} \tag{2.56}$$

As a result of this proportionality, R_L is a constant, $d\varphi/dx$ is a constant ($\varphi = \omega x/c$), $h = h_0$ is a constant and $B = B_0 \exp(-x/b)$. The condition that $d\varphi/dx$ is a constant implies that an observer travelling with the wave celerity ($x = ct$ and $\xi = \xi_0$) sees no change in both U and Z: $U = U(\xi_0)$ and $Z = Z(\xi_0)$. The function Ψ is determined by the downstream boundary condition. This implies that there are two equations with only one unknown function Φ. Hence the equations are dependent and there should be proportionality between the equations. Making use of the results from Equation 2.56 this yields:

$$\frac{\omega c}{-h\omega} = \frac{-g\omega}{\omega c} = \frac{-R_L b}{h} \tag{2.57}$$

This leads to the classical equation for the tidal wave propagation $c = \sqrt{(gh)}$ and also to the relationship between convergence and friction: $b = c/R_L$. The first equation is the same as the equation for the propagation of a disturbance in a relatively shallow water body, which also appears applicable in an ideal estuary, but which does not apply in estuaries where there is some degree of tidal damping or amplification. The second relationship is the condition for an ideal estuary, where the energy gained by convergence is balanced by the energy lost through friction. These two equations are special cases of the general solution derived from the non-linearized equations applicable to damped or amplified tidal waves, which is presented in Section 3.2.

Hence, Assumption 7, expressed in Equaitons 2.43 and 2.44, leads through the use of the linearized St. Venant equations to the Equations 2.41 and 2.42; or, in other words, Assumption 7 is only justified if the friction and the depth are constant and the width varies exponentially. In Section 3.2 it will be shown, also for the non-linear St. Venant equations, that if friction and convergence balance out ($b = k/c$), there is indeed no tidal damping and $c = \sqrt{(gh)}$.

Several other authors have made use of similar geometric conditions as in Equations 2.41–2.44, for example: Ketchum (1951) derived a theory based on a horizontal estuary bed; Abbott (1960) used a horizontal bed for the Thames and, Hunt (1964) used a constant depth and an exponentially varying width for the

Thames; Harleman (1966), in Ippen et al. (1966), used a constant depth and an exponential width variation for the Delaware.

McDowell and O'Connor (1977), elaborating on the concept of ideal estuaries developed by Pillsbury (1956), stated that since an ideal estuary implies that a unique relationship exists between maximum tidal discharges and channel cross-sectional area at all points along the estuary, this unique relationship might also exist between different estuaries of similar bed material. They analyzed the relationship between the size of tidal inlets and the maximum tidal flow, in much the same way as Bretting (1958) and Bruun and Gerritsen (1960) did.

In the following empirical assessment of the applicability of the shape of an ideal estuary it will be demonstrated that, although the assumptions required for an ideal estuary do not fully apply in most estuaries, the geometry of coastal plain estuaries generally can be described by Equations 2.41 and 2.42 and that, even if there is some bottom slope, Equation 2.38 always applies.

Empirical illustrations

We have already seen two examples of how real-life alluvial estuaries correspond with the shape of an ideal estuary. In Table 2.2 more examples of estuaries are given. The table provides values of a number of typical parameters that characterize alluvial estuaries. Besides data on shape (B, A, and h) there is information on the roughness (sometimes not more than estimates), the spring tidal range (H), the wave celerity (c), the tidal amplification (δ_H), and the length of the tidal intrusion (L_T).

In some estuaries, the longitudinal profile is split into two parts (mostly when there is a clear trumpet shape near to the estuary mouth). In those cases the logarithmic relation for A or B consists of two branches. For these estuaries a value of A'_0 is presented, which corresponds to the value that would have been obtained if the upstream branch were extended towards the estuary mouth. The observed rate of amplification δ_H, is defined as:

$$\delta_H = \frac{1}{H}\frac{\partial H}{\partial x} \tag{2.58}$$

The parameter R' is a friction parameter that will be presented in Section 3.1. The reason why it is presented here is because, if convergence ($1/a$) is larger than friction (R'), the tidal wave is amplified; if it is smaller, then it is damped. It can be seen in Table 2.2 that this indeed is the case. This aspect will be dealt with in detail in Section 3.1.

2.2.3 Assumptions on estuary shape in short estuaries

There are many alluvial estuaries that are not coastal plain estuaries, but which are forced by the underlying topography to be short. This occurs when the slope of the land is too steep for a long coastal plain to develop thereby forcing an

Table 2.2 Characteristic values of alluvial estuaries

Estuaries	A_0 (m^2)	A'_0 (m^2)	B_0 (m)	h (m)	a (km)	b (km)	C ($m/s^{0.5}$)	H (m)	c_0 (m/s)	δ_H ($10^{-6}m^{-1}$)	$1/b$ ($10^{-6}m^{-1}$)	R'/c ($10^{-6}m^{-1}$)	λ (km)	L_T (km)
Mae Klong	1400		250	5.2	102	155	53	2.0	7.1	−4.2	6.5	33.0	317	120
Limpopo	1710	1340	222	7.0	50	18	55	1.1	8.3	0.0	20.0	18.5	368	150
Lalang	2550		371	10.6	217	96	59	2.7	10.2	−1.0	4.6	8.7	453	200
Tha Chin	3000	1380	3600	5.3	87	87	45	2.6	3.0	−9.4	11.5	108.1	133	120
Sinnamary	3500	1210	2100	3.8	39	13	50	2.9	6.1	−5	25.6	66.0	271	
Chao Phya	4300		600	7.2	109	109	50	2.5	7.0	−3.6	9.2	26.8	311	120
Ord	7900		3200	4.0	22.1	15.2	50	5.9	6.3	0.0	45.2	33.7	278	65
Incomati	8100	1750	4500	2.9	42	42	60	1.4	3.6	−13.0	23.8	92.4	160	100
Pungué	28,000		6512	3.8	21	21	50	6.7	6.1	−8.5	47.6	253.2	271	120
Maputo	40,000	6460	9000	3.6	16	16	60	3.4	5.9	1.0	62.5	54.7	264	100
Thames	58,500		7480	7.1	23	23	55	4.3	8.3	2.3	43.5	19.7	371	110
Corantijn	69,000	34,600	30,000	6.5	64	48	55	2.3	8.0	−1.7	15.6	21.8	355	120
Gambia	84,400	27,200	9687	8.7	121	121	57	1.2	9.2	−1.0	8.3	12.4	410	500
Schelde	150,000		15,207	10.0	26	28	56	3.7	13.0	3.8	38.5	8.3	577	200
Delaware	255,000		37,655	6.6	41	42	55	1.5	8.0	1.7	24.4	20.8	357	200

estuary to be short. There are many of these estuaries in Great Britain, Australia, and the USA, although these countries also have several coastal plain estuaries (e.g. Thames, Delaware, Mississippi). It may be clear that mountainous islands generally have short estuaries as well. Prandle (2003), for instance, mainly describes this type of estuary and hence uses a different geometry than is done in this book. The article by Wright et al. (1973) is one of the few that specifically deals with short estuaries and also compares them to a coastal plain estuary. The interesting thing is that Wright et al. compare two branches of the same estuary system, the Ord and King rivers, which both are part of the Cambridge Gulf in the north of Western Australia. Another example of such a system is the Banyuasin–Lalang system on Sumatra, Indonesia). Here the Banyuasin is a short estuary and its main tributary, the Lalang, a coastal plain estuary. The Ord is a typical short estuary that fits the classification of Wright et al. (1973), namely:

1. a strong funnel shape with a cross-sectional area that obeys Equation 2.38;
2. the length of the tidal intrusion is finite, forced by the topography, and equal to $\lambda/4$;
3. a standing wave occurs, whereby the amplitude of the wave obeys:

$$\eta(x) = \eta_0 \cos\left(2\pi \frac{x}{\lambda}\right) \tag{2.59}$$

   where η_0 is the tidal amplitude at the estuary mouth.
4. there is no tidal damping or amplification on top of this, since there is a balance between friction and convergence.
5. both the depth and the width reduces exponentially.

The fourth item is motivated by the entropy principle, whereby there is uniform dissipation of energy, as we have seen in the previous section for coastal plain estuaries. To prove the validity of Equation 2.59, Wright et al. applied Green's law and field observations (Green, 1837), however, assumed frictionless flow and a progressive wave).

Several short estuaries, with a relatively small river discharge related to the tidal flood volume, appear to fit in this category. It will be shown in Section 3.1 that the assumption of a purely standing wave is at par with an un-amplified tidal wave. In fact a completely standing wave requires zero friction thereby introducing tidal amplification. The stronger the funnel shape, the greater the amplification. As this would obviously lead to an unsustainable tidal range if the estuary is long, there is a natural limit to this type of estuary at $x = \lambda/4$, i.e. at the first node of the standing wave where the tidal range is zero.

The condition for a purely standing wave that there is zero friction is not very workable in reality. There will always be friction and so the wave will not be a purely standing wave. This is also observed by Wright et al. (1973), who noticed that, although the wave travels very fast at almost infinite speed with slack occurring almost at the same time as HW and LW, there is a discernable time lag between the occurrence of HW along the estuary and even more so at LW. As a result the wave is not a purely standing wave (although ε is close to zero), there is friction, and there is a modest bottom level gradient (however, the convergence length of the depth is almost twice as large as the width convergence length, making the width convergence twice as pronounced as the shallowing).

Hence, although the theory is a bit more complex than described by Wright et al. (1973), their schematization is quite workable and also here we see that the schematization used for coastal plain estuaries, particularly Equation 2.38, may be applied.

2.3 RELATING TIDE TO SHAPE

2.3.1 Why look for relations between tide and shape?

Although ideal estuaries have a simple topography determined by: B_0, b, and h, the latter parameter is not always easy to determine. The width at the estuary mouth and the width convergence length can be readily measured from a map or aerial photograph, but the depth is much more complicated. With an echo sounder or a drop-weight, we can sound the depth at different locations. Determining the cross-sectional average depth along an estuary however is a lot of work. And there are other important tidal parameters that are not easy to measure, such as the tidal excursion E and the flood volume P_t.

To determine the estuary depth, an extensive hydrometric survey is required of cross-sectional areas at a number of locations. For reasonable accuracy, a minimum of ten cross sections over the salt intrusion length would be required. The tidal excursion can be measured directly by using floats, but such a method is cumbersome. It requires a full tidal cycle and has a relatively low accuracy due to wind effects and the non-uniform velocity distribution over the cross section. Also your floats tend to get stuck in the estuary bends. The flood volume is very difficult to measure directly requiring an extensive discharge measurement at the mouth of the estuary during a full tidal cycle. So let us see if we can derive analytical equations from the St. Venant equations, making use of the topography of an ideal estuary.

2.3.2 Theoretical derivations

Let us start with the flood volume P_t which is the integral of the tidal discharge between LWS and HWS at $x=0$:

$$P_t = \int_{\mathrm{LWS}}^{\mathrm{HWS}} Q(0,t)\mathrm{d}t \tag{2.60}$$

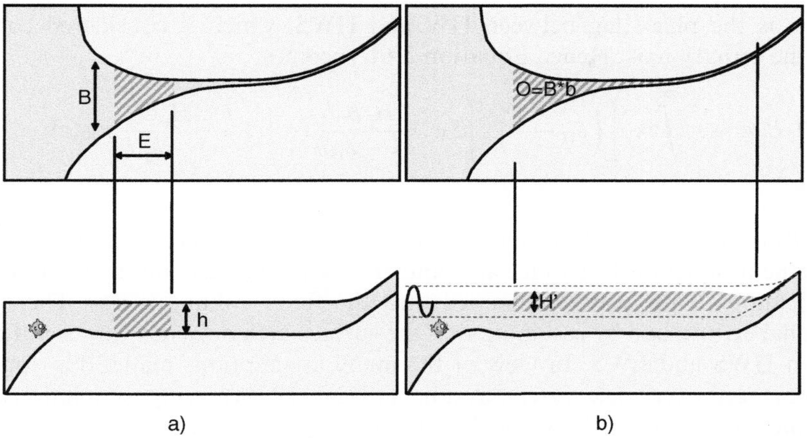

Figure 2.9 Definition sketch for the tidal flood volume: (a) as the product of B, h, and E; and (b) as the product of the surface area O and H', the tidal range between slacks.

where $Q(0,t)$ is the discharge at the estuary mouth. Since this integral is difficult to determine through direct measurement, it is approximated in two ways. The first is by equating the flood volume to the tidal prism enclosed by the high water line at HWS and the low water line at LWS in the estuary; the second, by equating it to the product of the cross-sectional area at the estuary mouth and the tidal excursion (see Figure 2.9). The first approach yields:

$$P_{\mathrm{t}} \approx \int_0^\infty H'B\mathrm{d}x \tag{2.61}$$

where $H'(x)$ is the range between HWS and LWS as a function of the distance: i.e. the difference between the envelopes of the water levels occurring at HWS and LWS along the estuary. Here, it is assumed that these levels are reached almost instantaneously along the estuary, implying that either the tidal wave is primarily standing ($\varepsilon \approx 0$) or that the wave length is large compared to the length of the estuary. In addition (for this derivation only), it is assumed that the tidal range is damped exponentially:

$$H = H_0 \exp(\delta_{\mathrm{H}} x) \tag{2.62}$$

This is the integral of Equation 2.58 for a constant rate of amplification, where H_0 is the tidal range at the estuary mouth and δ_{H} is the longitudinal rate of amplification of the tidal range (if $\delta_{\mathrm{H}} < 0$, the wave is damped). The relationship between H and H' can be seen simply from Figure 2.5 and reads:

$$H' = H\cos(\varepsilon) \tag{2.63}$$

where ε is the phase lag between HW and HWS, which is considered constant along the estuary axis. Hence Equation 2.61 becomes:

$$P_t \approx H_0 B_0 \cos(\varepsilon) \int_0^\infty \exp\left[\left(\delta_H - \frac{1}{b}\right)x\right] dx = \frac{H_0 B_0 b}{1 - \delta_H b}\cos(\varepsilon) = \frac{H_0 O}{1 - \delta_H b}\cos(\varepsilon) \tag{2.64}$$

where B_0 is the width at the estuary mouth and O is the surface area of the estuary. This is the product of the surface area and H', taking into account tidal damping or amplification. If the wave is damped then the flood volume is less. There is an additional error made by assuming that the surface area does not vary significantly between HWS and LWS. In view of the many assumptions made, this approach does not seem very accurate. Savenije (1992a, 1993a) using simulations of a hydraulic model showed however that the equation is accurate for a wide range of estuary shapes (depth ranging between 4 and 10 m; convergence length ranging between 10 and 100 km; tidal range ranging between 1.2 and 6 m).

The second approach yields:

$$P_t = \int_{\mathrm{LWS}}^{\mathrm{HWS}} A_0 U(0,t) dt \approx A_0 E_0 \tag{2.65}$$

where E_0 is the tidal excursion at the estuary mouth. The assumptions made here are: 1) that the cross-sectional area does not vary significantly with time; and 2) that the integral over time of the Eulerian velocity U between low water slack (LWS) and high water slack (HWS) is approximately equal to the tidal excursion E, whereas E is the integral between LWS and HWS of the Lagrangean velocity V of a moving water particle. The first assumption is acceptable in deep estuaries, the second is acceptable if the Froude number is small. To support this equation, Savenije (1992a, 1993a) demonstrated through model simulations that it is accurate as long as the tidal range to depth ratio is smaller than unity, implying that the Froude number is small.

Equating Equations 2.64 and 2.65 yields:

$$\frac{H_0 b}{(1 - \delta_H b)}\cos(\varepsilon) = h_0 E_0 \tag{2.66}$$

Essentially, this equation is a conservation of mass equation, where we have equated the volume change in the estuary between HWS and LWS to the amount of sea water entering the estuary during the same time.

Although this equation assumes a great deal, it is still a very interesting equation, providing a relationship between two parameters that we find difficult to measure directly: h_0 and E_0. An additional expression for the tidal excursion can be derived through Lagrangean analysis where we follow the water particle as it flows between LWS and HWS. Lagrangean analysis is also an adequate tool to develop Equation 2.66 more strictly.

Lagrangean analysis of a water particle

In a Lagrangean approach, the reference frame moves with the velocity of the water particle. The following equations apply:

$$V = \frac{\mathrm{d}x}{\mathrm{d}t} \tag{2.67}$$

$$S = \int_0^t V \mathrm{d}t \tag{2.68}$$

and hence:

$$x = x_0 + S \tag{2.69}$$

where $V = V(x,t)$ is the velocity of the moving particle and $S(x,t)$ is the distance travelled by the particle from a starting point x_0. From analysis of model simulations it appeared that the velocity variation in a Lagrangean reference frame was surprisingly regular and that it could be well described as a simple harmonic (Savenije, 1992b). If one assumes that the water particle moves according to a simple harmonic, starting at x_0 at LWS ($t=0$), then the velocity can be described by:

$$V(x,t) = \upsilon(x)\sin(\omega t) \tag{2.70}$$

$$\omega = 2\frac{\pi}{T} \tag{2.71}$$

where ω is the harmonic constant. In Equation 2.70, it is assumed that the influence of the fresh water discharge on the tidal velocities in the saline area is negligible.

In Equation 2.70, the effect of damping is present through the x-dependency of υ. Unlike the previous section the damping of the velocity amplitude is not necessarily exponential, meaning that the damping rate of the tidal velocity δ_U is not necessarily constant with x. We assume however that damping or amplification is modest ($|\delta_H b| \ll 1$), so that υ/H, or E/H is constant with x, according to Equation 2.66. Hence:

$$\delta_H = \delta_U = \frac{1}{\upsilon}\frac{\partial \upsilon}{\partial x} \tag{2.72}$$

where δ_U is the longitudinal relative rate of amplification of the velocity amplitude (which is negative in the case of damping). In the following derivations this damping is not neglected, but considered to be small over the distance travelled by the water particle ($|\delta_U E| \ll 1$).

The distance travelled by the water particle is found by substitution of Equation 2.70 in Equation 2.68 (for details see Intermezzo 2.3, where the subscript of δ_U has been dropped):

$$S = \frac{\upsilon}{\omega}(1 - \cos(\omega t)) \tag{2.73}$$

The tidal excursion E (the distance travelled between LWS and HWS) is obtained by substitution of $t = T/2$:

$$E = S(T/2) = \frac{\upsilon T}{\pi} = \frac{2\upsilon}{\omega} \tag{2.74}$$

which is the equation sought for the tidal excursion. Savenije (1992a, 1993a) demonstrated by model simulations that this equation underestimates the real value by 8 percent.

Intermezzo 2.3:

The integral of the Lagrangean velocity is the distance travelled S:

$$S = \int_0^t \upsilon \sin(\omega t)\mathrm{d}t = -\int_0^t \frac{\upsilon}{\omega}\mathrm{d}\cos(\omega t) = -\frac{\upsilon}{\omega}\cos(\omega t)\Big|_0^t + \frac{1}{\omega}\int_0^t \cos(\omega t)\mathrm{d}\upsilon$$

Since $\mathrm{d}\upsilon = \partial\upsilon/\partial x(V\mathrm{d}t) = \upsilon\delta V\mathrm{d}t = \upsilon^2\delta\sin(\omega t)\mathrm{d}t$, (where the subscript U of δ_U has been dropped) the second integral reads:

$$\frac{1}{\omega}\int_0^t \cos(\omega t)\mathrm{d}\upsilon = \frac{\upsilon^2\delta}{\omega}\int_0^t \cos(\omega t)\sin(\omega t)\mathrm{d}t = -\frac{\upsilon^2\delta}{4\omega^2}\cos(2\omega t)\Big|_0^t$$

Hence:

$$S = \frac{\upsilon}{\omega}\left(1 - \cos(\omega t) + \frac{\upsilon\delta}{4\omega}(1 - \cos(2\omega t))\right) = \frac{\upsilon}{\omega}\left(1 - \cos(\omega t) + \frac{E\delta}{8}(1 - \cos(2\omega t))\right)$$

Since $|\delta E| \ll 1$ in alluvial estuaries, the latter term may be disregarded. Similarly, differentiation of V with respect to t yields:

$$\frac{dV}{dt} = \omega \upsilon \cos(\omega t) + \frac{d\upsilon}{dt}\sin(\omega t)$$

The last term is further elaborated, considering that $\frac{\partial \upsilon}{\partial t} = 0$:

$$\frac{d\upsilon}{dt}\sin(\omega t) = \frac{\partial \upsilon}{\partial x} V \sin(\omega t) = \delta \upsilon^2 \sin^2(\omega t) = \omega \upsilon \frac{E\delta}{2}\sin^2(\omega t)$$

Hence:

$$\frac{dV}{dt} = \omega \upsilon \left(\cos(\omega t) + \frac{E\delta}{2}\sin^2(\omega t) \right) \approx \omega \upsilon \cos(\omega t)$$

since the term containing $\delta E/2$ may be disregarded. Note that in these derivations δ may still be a function of x.

In a Eulerian reference frame the harmonics are distorted. Through the introduction of the argument ξ (similar to Equation 2.50) the above equations can be transformed into a Eulerian reference frame:

$$\xi = \omega t - \frac{\omega(x - x_0 - S)}{c} \tag{2.75}$$

yielding:

$$U = \upsilon \sin(\xi) \tag{2.76}$$

where $U = U(x,t)$ is the velocity of flow at a certain location at a certain time and c is the tidal wave celerity. If we move with the water particle, then $x = x_0 + S$, $\xi = \omega t$, and $U = V$. At $x = x_0 + E$, a time lag of E/c occurs between HWS of the moving particle and HWS observed at $x = x_0$ (see Figure 2.10).

With this transformation, the Eulerian continuity equation for one-dimensional flow, Equation 2.27, can be transformed into the Lagrangean reference frame, where it can be solved analytically. Combination of Equation 2.27 with Equation 2.39 yields:

$$r_S \frac{\partial h}{\partial t} + U\frac{\partial h}{\partial x} + h\frac{\partial U}{\partial x} - \frac{hU}{b} = 0 \tag{2.77}$$

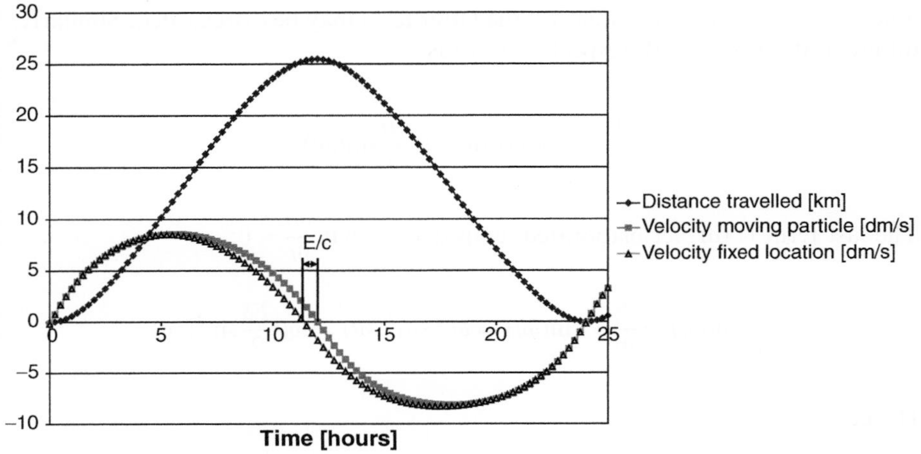

Figure 2.10 Velocity (dm/s) and distance travelled (km) by a water particle.

Partial differentiation of Equation 2.76 with respect to x for a moving particle where $x = x_0 + S$, and combination of the result with Equations 2.72 and 2.75 yields a Lagrangean expression for the partial derivative of U with respect to x:

$$\frac{\partial U}{\partial x} = -\frac{1}{c}\frac{\mathrm{d}V}{\mathrm{d}t} + \delta_{\mathrm{U}} V \tag{2.78}$$

the details of which are explained in Intermezzo 2.4:

Intermezzo 2.4:

$$\frac{\partial U}{\partial x} = \frac{\partial \upsilon}{\partial x}\sin\xi + \upsilon\cos\xi\frac{\partial \xi}{\partial x} = \delta_{\mathrm{U}}\, V + \frac{1}{\omega}\frac{\mathrm{d}V}{\mathrm{d}t}\frac{\partial \xi}{\partial x}$$

$$\frac{\partial \xi}{\partial x} = -\frac{\omega}{c}\left(1 - \frac{\partial S}{\partial x}\right) = -\frac{\omega}{c}\left(1 - \frac{\partial \upsilon}{\partial x}\frac{1}{\omega}(1 - \cos(\omega t))\right) = -\frac{\omega}{c}(1 - \delta_{\mathrm{U}}\, S)$$

Substitution of $\partial\xi/\partial x$ and $\mathrm{d}V/\mathrm{d}t$ (from Intermezzo 2.3) in the equation for $\partial U/\partial x$ yields:

$$\frac{\partial U}{\partial x} = \delta_U V - \frac{1}{c}\frac{\mathrm{d}V}{\mathrm{d}t}(1 - \delta_{\mathrm{U}}\, S)$$

Equation 2.78 is obtained under the assumption that $|\delta_U S| < |\delta_U E| \ll 1$. Note that in this derivation δ may still be a function of x.

Moreover, the variation of the water depth with time for a moving water particle is described by:

$$\frac{dh}{dt} = \frac{\partial h}{\partial t} + V\frac{\partial h}{\partial x} \tag{2.79}$$

Substitution of Equations 2.78 and 2.79 into Equation 2.77 yields the continuity equation for a moving volume of water ($U=V$) in a Lagrangean reference frame. Here, it has been assumed that 1) r_S is close to unity and 2) that the Froude number is small. As a result, the second term of Equation 2.79 is much smaller than the first. Therefore, the introduction of the storage width ratio in the second term creates only a third-order error. Hence:

$$r_S\frac{dh}{dt} = +\frac{h}{c}\frac{dV}{dt} + hV\frac{(1-\delta_U b)}{b} \tag{2.80}$$

Elaboration of Equation 2.80 yields:

$$r_S\frac{dh}{h} = \frac{(1-\delta_U b)}{b}dx + \frac{1}{c}dV \tag{2.81}$$

where the first term of the right-hand member is a conservation of mass term leading to depth gain as a result of the convergence of the banks and deceleration due to damping, which the moving particle experiences. The second term is a conservation of mass term that drives water level variation as a result of the velocity variation. Integration between LWS ($t=0$) and t yields:

$$r_S \ln(h) = r_S \ln(h_{LWS}) + S\frac{(1-\delta_U b)}{b} + \frac{V}{c} \tag{2.82}$$

When plotted against the distance travelled, this is an exponential function of an inclined ellipse (see Figure 2.11). The drawn curve through the heart of the inclined ellipse represents the second term on the right-hand side. If $b \to \infty$ this term is zero (the case of a constant cross section and no damping) and the inclination of the ellipse disappears. In this case the velocity and water depth near the moving particle are in phase creating an undamped progressive wave. This is in agreement with other analytical methods where a progressive wave occurs in a channel of constant cross section, of infinite length, and without friction.

Now it looks as if in the derivation of Equation 2.82, the damping rate δ_U has been assumed to be constant with x, but this is not the case. In Intermezzo 2.5 it

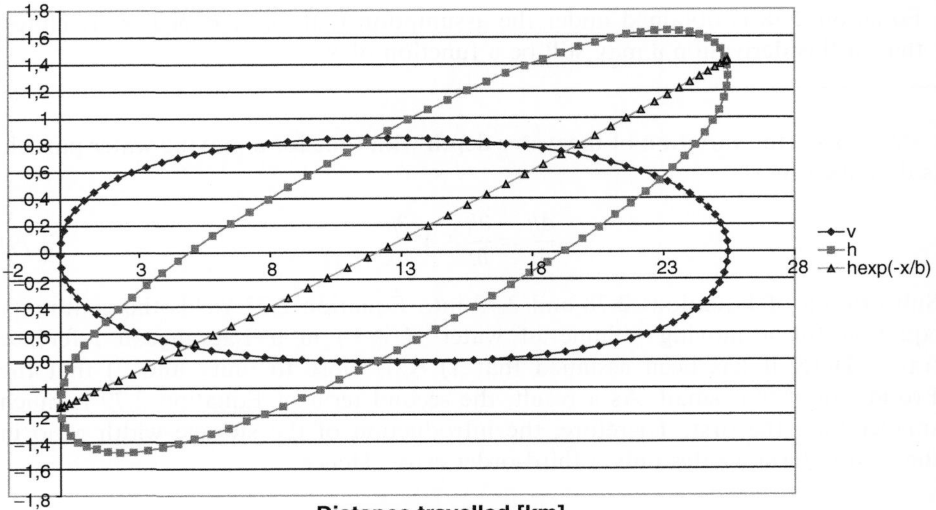

Figure 2.11 Velocity (m/s) and elevation (m) of a moving water particle as a function of the distance travelled (km).

is demonstrated that Equation 2.82 follows from Equation 2.81 using the solution for the tidal damping rate derived in Section 3.1.2, which is not constant with x.

Intermezzo 2.5:

Rearrangement of Equation 2.81 and using $\delta = \delta_U = \delta_H$ yields:

$$r_S \frac{dh}{h} = \frac{1}{b}dx + \frac{1}{c}dV - \delta dx = \frac{dx}{b} + \frac{dV}{c} - \frac{d\upsilon}{\upsilon} = \frac{dx}{b} + \frac{dV}{c} - \frac{dH}{H}$$

Using the dimensionless tidal range $y = H/H_0$, integration from LWS to a certain point in time yields:

$$r_S \ln\left(\frac{h}{h_{LWS}}\right) = \frac{S}{b} + \frac{V}{c} - (\ln y - \ln y_{LWS})$$

Now making use of Equation 3.25, this equation can be written as:

$$r_S \ln\left(\frac{h}{h_{LWS}}\right) = \frac{S}{b} + \frac{V}{c} - \frac{S}{\beta}\left(1 - \frac{\beta}{\alpha}\frac{y - y_{LWS}}{S}\right) \approx \frac{S}{b} + \frac{V}{c} - \frac{S}{\beta}\left(1 - \frac{\beta}{\alpha}\frac{dy}{dx}\right)$$

This last step is a first-order approximation which is correct as long as $|\delta S| \ll 1$. Making use of Equation 3.23, this modifies into:

$$r_S \ln\left(\frac{h}{h_{LWS}}\right) \approx \frac{S}{b} + \frac{V}{c} - \frac{S}{\beta}\left(1 - \frac{y}{\alpha + y}\right) = \frac{S}{b} + \frac{V}{c} - \frac{S}{\beta}\left(\frac{\alpha}{\alpha + y}\right) = \frac{S}{b} + \frac{V}{c} - S\delta$$

which is the same as Equation 2.82.

A relation between the estuary depth and the tidal range

Equating the left-hand side of Equation 2.81 to zero, allows the determination of the times at which high water (HW) and low water (LW) occur. Subsequent integration yields:

$$t_{HW} = \frac{T}{2} - \frac{1}{\omega}\arctan\left(\frac{\omega}{c}\frac{b}{(1 - \delta b)}\right) \tag{2.83}$$

$$t_{LW} = T - \frac{1}{\omega}\arctan\left(\frac{\omega}{c}\frac{b}{(1 - \delta b)}\right) \tag{2.84}$$

where we have dropped the subscript for δ.

Substitution of these times in V (yielding V_{HW} and V_{LW}) and in S (yielding S_{HW} and S_{LW}), substitution of these results in Equation 2.82 (yielding $\ln(h_{HW})$ and $\ln(h_{LW})$) and subsequent subtraction of $\ln(h_{HW})$ and $\ln(h_{LW})$ yields an expression from which the tidal range of the moving particle can be determined:

$$\begin{aligned} r_S(\ln(h_{HW}) - \ln(h_{LW})) &= \frac{(1 - \delta b)}{b}(S_{HW} - S_{LW}) + \frac{1}{c}(V_{HW} - V_{LW}) \\ &= \frac{(1 - \delta b)}{b}(2S_{HW} - E) + 2\frac{V_{HW}}{c} \end{aligned} \tag{2.85}$$

The second step of Equation 2.85 follows from the symmetry of the periodic functions: $V_{HW} = -V_{LW}$ and $S_{HW} + S_{LW} = E$ (see Figure 2.11). Equation 2.85 is an analytical relation between maximum and minimum water level on the one hand, and tidal velocity amplitude υ on the other, given a certain periodicity, estuary shape, and mean estuary depth. The expressions for S_{HW} and V_{HW} can be found by substitution of Equation 2.83 into Equations 2.70 and 2.73:

$$S_{HW} = \frac{\upsilon}{\omega}(1 + \cos(\varepsilon)) = E\frac{1 + \cos(\varepsilon)}{2} \tag{2.86}$$

$$V_{HW} = \upsilon \sin(\varepsilon) \tag{2.87}$$

where:

$$\varepsilon = \arctan\left(\frac{\omega b}{c(1-\delta b)}\right) \tag{2.88}$$

A result of the definition of the arc-tangent is that $-\pi/2 < \varepsilon < \pi/2$. For damped or moderately amplified tidal waves the argument of the arc-tangent is positive ($1-\delta b \geq 0$), therefore $0 \leq \varepsilon < \pi/2$. Substitution of Equations 2.86 and 2.87 in Equation 2.85 and some goniometric manipulation yields:

$$\ln(h_{HW}) - \ln(h_{LW}) = \frac{E}{r_S b}\frac{(1-\delta b)}{\cos(\varepsilon)} \tag{2.89}$$

or:

$$h_{HW} = h_{LW}\exp\left(\frac{E}{r_S b}\frac{(1-\delta b)}{\cos(\varepsilon)}\right) \tag{2.89}$$

Savenije (1992a, 1993a) demonstrated that by using a Taylor-series expansion of the exponential function in Equation 2.89, this equation can be transformed into Equation 2.90. This occurs only when the argument of the exponential function in Equation 2.89 is smaller than unity. In all practical cases $E/b \ll 1$ and damping is modest ($\delta b < 1$); hence, the approximation is correct.

$$\frac{H}{E} = \frac{\eta\omega}{\upsilon} = \frac{\bar{h}}{r_S b}\frac{(1-\delta b)}{\cos(\varepsilon)} \tag{2.90}$$

or:

$$\frac{H}{\bar{h}} = \frac{E}{r_S b}\frac{(1-\delta b)}{\cos(\varepsilon)} \tag{2.91}$$

where $\bar{h}$ is the tidal average estuary depth (not necessarily constant along the estuary). In the text, for reasons of convenience, we shall drop the over-bar and simply use h to describe the tidal average depth. If the tidal average depth is constant along the estuary we shall use h_0.

Equation 2.90 is essentially the same as Equation 2.66 which we derived intuitively from the water balance equation. Now that the equation has been formally derived, it provides an accurate relation between H/E and h/b. This is the analytical relation we were looking for to relate the depth to the tidal excursion. We shall call it the 'Geometry-Tide relation' since it presents a direct relation between the tidal scales H and E with the geometric scales h and b. It is good to realize that the ratio of the vertical to the horizontal tidal range (H/E) is directly proportional to the ratio of the vertical to the horizontal scales of the estuary shape (h/b).

An interesting result is obtained if we combine Equations 2.88 and 2.91:

$$\frac{\eta}{h} = \frac{\upsilon}{c} \frac{1}{r_S \sin(\varepsilon)} \tag{2.92}$$

which is a direct relationship between the amplitude-to-depth ratio, the Froude number, and the Wave-type number, $N_E = \sin(\varepsilon)$. This very useful equation provides a relationship between important hydraulic scales and is therefore named the 'Scaling equation.'

It is interesting to note that a simpler version of this equation is widely used in the literature, particularly in perturbation analysis (e.g. Jay, 1991; Friedrichs and Aubrey, 1994). In perturbation analysis it is customary to neglect $\sin(\varepsilon)$ in Equation 2.92. This assumes $\varepsilon = \pi/2$, namely the case of a purely progressive wave. Hence in perturbation analysis it is implicitly assumed that the wave behaves as a progressive wave, whereas we know that the tidal wave in an estuary is of mixed character (see Figure 2.5). Because ε is small (typically in the order of 0.3) the error made by neglecting $\sin(\varepsilon)$ is substantial. In the author's experience, the Froude number is always substantially smaller than the tidal amplitude to depth ratio. It is an illustration of how perturbation analysis can easily introduce unnecessary errors. While Equation 2.92 is no more complicated, it is certainly more accurate than the equation used in perturbation analysis.

Friedrichs and Aubrey (1994) also present a version of Equation 2.90 obtained from perturbation analysis, but in that equation r_s, $\cos(\varepsilon)$, and $(1 - \delta b)$ are missing. In fact, the disregard of $\cos(\varepsilon)$ is the same as assuming that the tidal wave behaves like a standing wave. Because ε is small, the error may not seem so large, but combined with the neglect of damping and the effect of storage width, it can result in considerable deviation from the true value. A deviation which can be avoided by more rigorous analytical derivation.

Tidal Damping

We can conclude from Equation 2.90 that, for the ratio of H/E to be independent of x, damping should be small over the length of the estuary ($|\delta_U b| \ll 1$). In such a case, the right-hand side of Equation 2.90 is constant with x. As a consequence, the derivative of H/E with respect to x is zero, resulting in:

$$\frac{1}{E}\frac{\partial E}{\partial x} \approx \frac{1}{H}\frac{\partial H}{\partial x} \text{ if } |\delta b| \ll 1 \tag{2.93}$$

and since there is a linear relation between υ and E:

$$\delta_{\mathrm{U}} = \frac{1}{\upsilon}\frac{\partial \upsilon}{\partial x} \approx \delta_{\mathrm{H}} = \frac{1}{H}\frac{\partial H}{\partial x} \text{if } |\delta b| \ll 1 \tag{2.94}$$

This is a useful formula as, in practice, δ_{H} is much easier to determine than δ_{U}. Therefore we shall make no further distinction between these amplification factors and drop the subscript.

3

Tidal dynamics

In the previous chapter, we derived relations between the hydraulic parameters of estuary flow and the topography. In this chapter, we shall derive analytical equations for tidal damping/amplification (on the basis of the conservation of momentum equation) and wave celerity (using the combined mass and momentum equations). These equations are derived through the method of characteristics and Lagrangean analysis. This approach uses analytical derivation and not scaling. The latter, called perturbation analysis, is useful for identifying the main mechanisms at play and for assessing orders of magnitude, but, used as a tool for derivation, does not always result in correct equations, as will be demonstrated. The analytical equations derived in this chapter are more general versions—or refinements—of well-known (classical) equations, such as Green's law and other rules of thumb. Most of these classical equations are only correct for frictionless channels with a constant topography, or apply to either progressive or standing waves. The general equations derived in this book apply to the full range of tidal waves (with a phase lag varying between 0 and $\pi/2$) and the natural topographies of alluvial estuaries.

3.1 TIDAL MOVEMENT AND AMPLIFICATION

3.1.1 Why is the tidal wave amplified or damped?

We saw in the previous chapter that tidal amplification (or damping) has an effect on the water balance equation, and therefore on the ratio of E/H and (importantly) on the wave celerity (see Equation 2.88). But what causes a tidal wave to be amplified or damped? Until now, we have only analyzed the mass balance equation, although the assumed Lagrangean velocity function is a solution of both St. Venant equations. To understand the reasons for tidal damping, we have to look at the momentum balance equation.

Early authors like Langbein (1963) and Dyer (1973) suggested that tidal amplification is the result of the imbalance between topographic convergence and friction. If convergence is stronger than friction, the wave is amplified; if friction is stronger than convergence, the wave is damped; if their impact is equal, the tidal range is constant. This was indeed demonstrated by Jay (1991), who used perturbation theory, and by Savenije (1998) who combined the Lagrangean equations of the previous chapter with the momentum balance equation. In the following section we shall derive a relation for tidal amplification or damping.

3.1.2 Derivation of the tidal damping equation

We can rearrange Equation 2.80 as follows:

$$\frac{\mathrm{d}V}{\mathrm{d}t} = r_S \frac{c}{h}\frac{\mathrm{d}h}{\mathrm{d}t} - \frac{cV}{b} + \frac{cV}{H}\frac{\mathrm{d}H}{\mathrm{d}x} \tag{3.1}$$

where δ has been replaced by Equation 2.58, and δ is not necessarily constant.

The momentum balance equation Equation 2.23, when written in a Lagrangean reference frame, yields:

$$\frac{\mathrm{d}V}{\mathrm{d}t} + g\frac{\partial h}{\partial x} + g(I_b - I_r) + g\frac{V|V|}{C^2 h} = 0 \tag{3.2}$$

Combination of Equations 3.1 and 3.2, and using Equation 2.67 yields:

$$r_S \frac{cV}{gh}\frac{\mathrm{d}h}{\mathrm{d}x} - \frac{cV}{g}\left(\frac{1}{b} - \frac{1}{H}\frac{\mathrm{d}H}{\mathrm{d}x}\right) + \frac{\partial h}{\partial x} + I_b - I_r + \frac{V|V|}{C^2 h} = 0 \tag{3.3}$$

Moreover, because the tidal range H is the difference between h_{HW} and h_{LW}:

$$\frac{\mathrm{d}H}{\mathrm{d}x} = \frac{\mathrm{d}h_{HW}}{\mathrm{d}x} - \frac{\mathrm{d}h_{LW}}{\mathrm{d}x} \tag{3.4}$$

Both for HW and LW, by definition:

$$\frac{\partial h}{\partial t} = 0 \tag{3.5}$$

and hence:

$$\frac{\mathrm{d}h}{\mathrm{d}x} = \frac{\partial h}{\partial x} \tag{3.6}$$

If the dimensionless tidal wave (scaled by the tidal range) is considered un-deformed (which is the case when $H/h \ll 1$), the damping is symmetrical in respect of the average water level, which has a slope I:

$$\frac{\mathrm{d}h_{HW}}{\mathrm{d}x} + \frac{\mathrm{d}h_{LW}}{\mathrm{d}x} \approx I \tag{3.7}$$

In addition:

$$h_{HW} \approx \bar{h} + \eta \tag{3.8}$$

$$h_{HW} \approx \bar{h} - \eta \tag{3.9}$$

and:

$$V_{\mathrm{HW}} = \upsilon \sin \varepsilon \tag{3.10}$$

$$V_{\mathrm{LW}} = -\upsilon \sin \varepsilon \tag{3.11}$$

where $\eta = H/2$. Combination of Equations 3.1, 3.6, 3.8, and 3.10 yields for HW:

$$\frac{r_S c \upsilon \sin \varepsilon}{g(\bar{h}+\eta)} \frac{\mathrm{d}h_{\mathrm{HW}}}{\mathrm{d}x} - \frac{c\upsilon \sin \varepsilon}{g}\left(\frac{1}{b} - \frac{1}{H}\frac{\mathrm{d}H}{\mathrm{d}x}\right) + \frac{\mathrm{d}h_{\mathrm{HW}}}{\mathrm{d}x} + \frac{(\upsilon \sin \varepsilon)^2}{C^2(\bar{h}+\eta)} = -I_{\mathrm{b}} + I_{\mathrm{r}} \tag{3.12}$$

Similarly for LW:

$$\frac{-r_S c \upsilon \sin \varepsilon}{g(\bar{h}-\eta)} \frac{\mathrm{d}h_{\mathrm{LW}}}{\mathrm{d}x} + \frac{c\upsilon \sin \varepsilon}{g}\left(\frac{1}{b} - \frac{1}{H}\frac{\mathrm{d}H}{\mathrm{d}x}\right) + \frac{\mathrm{d}h_{\mathrm{LW}}}{\mathrm{d}x} - \frac{(\upsilon \sin \varepsilon)^2}{C^2(\bar{h}-\eta)} = -I_{\mathrm{b}} + I_{\mathrm{r}} \tag{3.13}$$

Subtraction of 3.12 and 3.13 yields:

$$\frac{r_S c \upsilon \sin \varepsilon}{\bar{h}+\eta}\left(\frac{\mathrm{d}h_{\mathrm{HW}}}{\mathrm{d}x} + \frac{\mathrm{d}h_{\mathrm{LW}}}{\mathrm{d}x}\frac{(\bar{h}+\eta)}{(\bar{h}-\eta)}\right) - \frac{2c\upsilon \sin \varepsilon}{\bar{h}}\left(\frac{\bar{h}}{b} - \frac{\bar{h}}{H}\frac{\mathrm{d}H}{\mathrm{d}x}\right) + g\frac{\mathrm{d}H}{\mathrm{d}x} + 2f'\frac{(\upsilon \sin \varepsilon)^2}{\bar{h}} = 0 \tag{3.14}$$

with:

$$f' = \frac{g}{C^2}\left(1 - \left(\frac{\eta}{\bar{h}}\right)^2\right)^{-1} = f\left(1 - \left(\frac{\eta}{\bar{h}}\right)^2\right)^{-1} \tag{3.15}$$

where f' is the adjusted friction factor taking account of the friction being larger at LW than at HW. If the tidal amplitude-to-depth ratio is small, $f' \approx f$.

The parameters between parentheses in the first term of Equation 3.14 can be replaced by I of Equation 3.7, provided $\eta/h < 1$. Elaboration yields:

$$\frac{\bar{h}}{H}\frac{\mathrm{d}H}{\mathrm{d}x}\left(1 + \frac{gH}{2c\upsilon \sin \varepsilon}\right) = \frac{\bar{h}}{b} - f'\frac{\upsilon \sin \varepsilon}{c} - \frac{r_S I}{(2 + H/\bar{h})} \tag{3.16}$$

In the following text, for convenience sake, we shall drop the over bar and h will stand for the tidal average depth.

Now let us consider the order of magnitude of the last term compared to h/b. Over the distance L of the tidal influence the bottom slope is negligible, hence $I \ll h/L$. Moreover, in all estuaries b is several times smaller than L (see Table 2.2).

As a result, I is at least an order of magnitude smaller than h/b during periods of low river discharge. Therefore, the last term can be disregarded in most practical cases. Further research into the relative importance of I versus h/b may be needed (particularly in the upstream part of an estuary), which is complicated by the fact that I is difficult to observe accurately.

Hence, the analytical solution of the St. Venant's equations yields:

$$\frac{\bar{h}}{H}\frac{\mathrm{d}H}{\mathrm{d}x}\left(1+\frac{g\eta}{c\upsilon\sin\varepsilon}\right)=\frac{\bar{h}}{b}-f'\frac{\upsilon\sin\varepsilon}{c} \tag{3.17}$$

Subsequently, let us analyze the origin of the terms in this equation. The first term on the right hand side is obviously the convergence term of the continuity equation. The second term stems from the friction term in the momentum balance equation. On the left hand side, it is less obvious. The '1' stems from the last term in Equation 3.1 and is the term that determines the effect of tidal damping on the mass balance equation; the second term between brackets stems from the depth gradient in the momentum balance equation. Scaling (see below) shows that this term is larger than 1.

This equation is a general version of Green's law, a rule of thumb often quoted. Green (1837) assumed that the amount of energy in a progressive tidal wave ($E=0.5\rho g\eta^2 BcT$) would remain constant under frictionless flow as it travels up a converging estuary. If we use the classical equation for wave propagation ($c^2=gh$), this leads to the tidal range being inversely proportional to the square root of the width and the 0.25th power of the depth. In an ideal estuary with constant depth, it implies that $\delta_H=1/(2b)$. If indeed we consider zero friction ($f'=0$), a progressive wave ($\sin\varepsilon=1$), and $r_S=1$, then combination of Equation 3.17 with 2.92 leads indeed to $\delta_H=1/(2b)$.

From Equation 3.17, it can be seen that in an ideal estuary where there is no tidal damping or amplification:

$$\frac{1}{b}=f'\frac{\upsilon\sin\varepsilon}{\bar{h}c}=\frac{R'}{c} \tag{3.18}$$

This is the same result as in Equation 2.57, which was the condition for an ideal estuary to occur. The resistance term R'/c is also presented in Table 2.2. We can indeed verify the earlier remark that there is tidal amplification of $1/b>R'/c$, see Figure 3.1. We can also verify that since b, f, h, and υ are constant along the estuary, the wave celerity and $\sin(\varepsilon)$ are proportional. Since $\sin(\varepsilon)$ indicates the type of tidal wave (it equals zero for a standing wave and 1 for a progressive wave) it is called the Wave-type number N_E (Savenije, 1998). In alluvial estuaries, where the tidal wave is of a mixed character, the Wave-type number is between 0 and 1. Since it has been observed that the phase lag is constant along an estuary, the wave

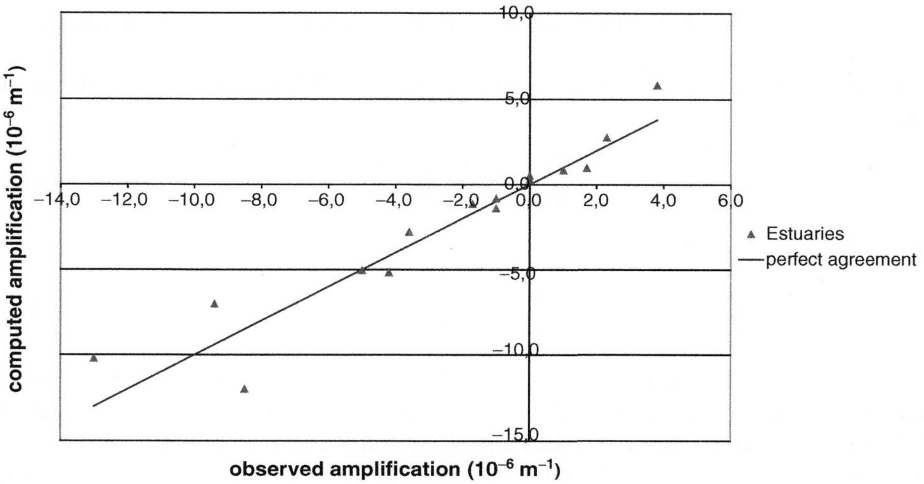

Figure 3.1 Computed and observed tidal damping and amplification (δ_H) in the estuaries of Table 2.2.

celerity also is, which can indeed be observed in estuaries, at least for considerable stretches where convergence and depth are constant.

To simplify Equation 3.18, three parameters are introduced, the dimensionless tidal range y, the dimensionless Tidal Froude number α, and the tidal damping scale β which has a length dimension:

$$y = \frac{H}{H_0} \tag{3.19}$$

$$\alpha = \frac{2cv \sin \varepsilon}{gH_0} = \frac{cv \sin \varepsilon}{g\eta_0} \tag{3.20}$$

$$\frac{1}{\beta} = \frac{1}{b} - f' \frac{v \sin \varepsilon}{c\bar{h}} \tag{3.21}$$

where H_0 is the tidal range at $x = 0$. The Tidal Froude number α differs from the regular Froude number in that it contains the tidal amplitude-to-depth ratio (η/h) and $\sin(\varepsilon)$. We can assess the order of magnitude of α by substitution of the Scaling equation (Equation 2.92). Considering that $r_s c^2 \approx gh$, α scales at $\sin^2\varepsilon$, which is O(0.1).

The tidal damping scale β is the length scale of tidal damping, which is negative in the case of tidal damping (when friction outweighs amplification due to the conversion of the banks) and positive in the case of tidal amplification. The two parameters α and β contain parameters which according to earlier studies

(Savenije, 1992a,b, 1993, 1998) may be considered constant along the longitudinal axis of the estuary.

Substitution of y, α, and β into Equation 3.17 leads to the following differential equation:

$$\frac{\mathrm{d}y}{\mathrm{d}x}\left(1+\frac{1}{\alpha}y\right)=\frac{1}{\beta}y \tag{3.22}$$

or

$$\frac{\mathrm{d}y}{\mathrm{d}x}=\frac{\alpha}{\beta}\frac{y}{(\alpha+y)} \tag{3.23}$$

It can be seen from Equation 3.23 that when $\alpha \gg y$, the tidal range increases or decreases exponentially with x at a length scale equal to β. When $\alpha \ll y$ however, then $\mathrm{d}y/\mathrm{d}x$ is constant and damping or amplification is linear with angular coefficient α/β. Where y approaches zero as a result of tidal damping, the damping becomes exponential (even if α is small) preventing y from intersecting with the x-axis. It can also be seen that if β approaches infinity, the tidal range is constant. Hence, the magnitude of damping or amplification depends on the value of α/β. If α/β is close to 0, damping is modest or linear. If α/β is large, negatively or positively, damping or amplification is exponential.

Equation 3.23 can be integrated by the separation of variables (personal communication by Veling):

$$\mathrm{d}x=\left(\frac{\beta}{y}+\frac{\beta}{\alpha}y\right)\mathrm{d}y \tag{3.24}$$

With $y=1$ ($H=H_0$) at $x=0$, integration yields:

$$x=\beta\ln(y)+\frac{\beta}{\alpha}(y-1) \tag{3.25}$$

This is a very simple relation consisting of a logarithmic and a linear term. The only complication in Equation 3.25 is that the dependent variable y cannot be written as an explicit function of x. Figures 3.2 and 3.3 present plots of this relation (for positive and negative values of β/α), where y and x are plotted on the horizontal and vertical axes, respectively. When $\alpha/\beta=0$ ($\beta/\alpha\to\infty$), the ideal estuary occurs (as a special case) with no tidal damping ($y=1$).

Although Equation 3.25 cannot be written as $y=\mathrm{f}(x)$, one can see that if β/α is large, the first term becomes less important and Equation 3.25 is closely reproduced by a simple line:

$$y=\frac{H}{H_0}=1+\frac{\alpha}{\beta}x \tag{3.26}$$

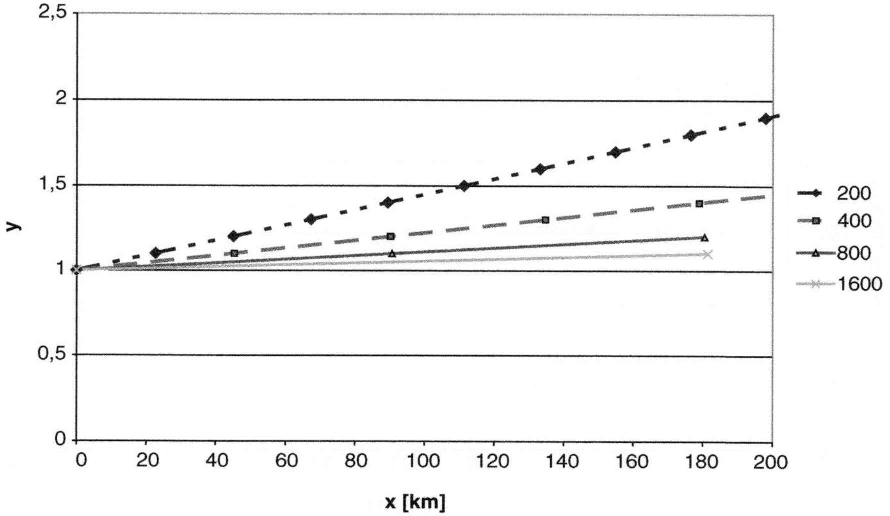

Figure 3.2 Plots of Equation 3.25 for tidal amplification (positive values of β/α).

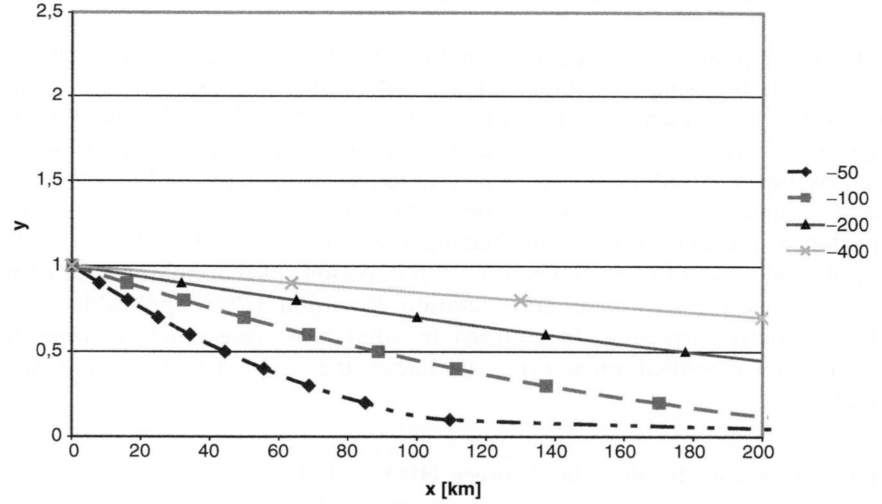

Figure 3.3 Plots of Equation 3.25 for tidal damping (negative values of β/α).

It appears that in natural estuaries, β/α is large. In fact, β/α is the length scale of the linear damping (or amplification) process, whereas β is the length scale of the exponential process. Since $\alpha<1$ in all alluvial estuaries, tidal amplification and damping are predominantly linear (except close to the point where the tidal wave dies out). Figure 3.4 shows a plot of the values of $1/\beta$ and $1/\alpha$ obtained from Table 2.2. In all simulated cases, $\beta/\alpha > 130$ km for amplification and $\beta/\alpha < -75$ km

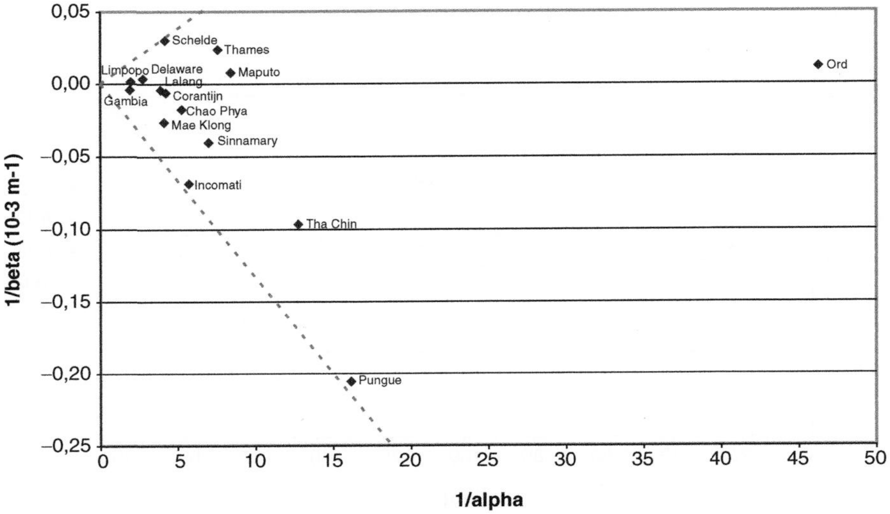

Figure 3.4 Values of $1/\alpha$ and $1/\beta$ for the estuaries of Table 2.2. The plots are contained between the lines $\beta/\alpha = 150$ km and $\beta/\alpha = -75$ km.

for tidal damping (see dashed lines in Figure 3.4). With these values, the error incorporated by using Equation 3.26 instead of Equation 3.25 is not so large. Equation 3.26 represents a bundle of lines through the point (0,1) where the tidal range $H = H_0$ at $x = 0$. The exponential term, $\beta \ln(y)$, only becomes important when y approaches zero and, hence, when α is no longer small compared to y. When the line of Equation 3.26 approaches the x-axis the exponential term, which gains importance, prevents it from intersecting the x-axis. Thus Equation 3.26 loses its applicability near the point where $x = -\beta/\alpha$. As long as that point lies far enough upstream from the area under consideration, Equation 3.26 is applicable.

The result is surprising. In contrast to what most researchers assume, tidal damping and amplification is primarily linear; the simplest relation that can be thought of.

The importance of the phase lag between HWS and HW

In the relative importance of the two terms of Equation 3.25, the Wave-type number ($N_E = \sin\varepsilon$) plays a key role. A purely progressive wave, where $\sin\varepsilon = 1$, only occurs in prismatic channels (where b tends to infinity) and where β is negative and dominated by the friction term. Consequently, a progressive wave has a relatively large value of α. The result is exponential damping, which one normally observes in rivers and canals where progressive waves occur. A purely standing wave occurs when $\varepsilon = 0$. As a result, α is relatively small, $\beta = b$ and tidal amplification is linear although approaching zero, as was observed by Wright et al. (1973) and illustrated by the Ord case in Table 2.2.

In estuaries, it appears that ε prevents the development of exponential amplification through an implicit feedback mechanism. In estuaries where strong tidal amplification occurs, the value of ε becomes small thereby decreasing the value of α, so that, in turn, the exponential amplification is suppressed. Figure 3.4 shows that for a β value of $10\,\mathrm{km}$ $(1/\beta = 0.1*10^{-3})$ which would otherwise imply strong amplification, α assumes a value of less than 7 percent $(1/\alpha > 15)$. The ratio β/α then equals $150\,\mathrm{km}$, which is a large length scale for amplification. The reduction of α through the Wave-type number therefore largely decreases the amplification.

Although the phase lag is generally disregarded in perturbation analysis, it was recognized early by Van Veen (1937, 1950) to be a crucial parameter in tidal hydraulics. This researcher, who published in Dutch, presented a formula similar to Equation 2.64.

3.1.3 Application of the derived formula to observations

We already saw in Figure 3.1 that Equation 3.17 fits very well with the observed values. But, these are average values over a reach of the estuary and do not show if the longitudinal variation of the tidal range also fits observations. In Savenije (2001a), several applications of Equation 3.23 to observations in real estuaries were presented. Here, for the purpose of illustration, only examples of the Schelde and Incomati are given in Figures 3.5 and 3.6, respectively. In the Schelde, the

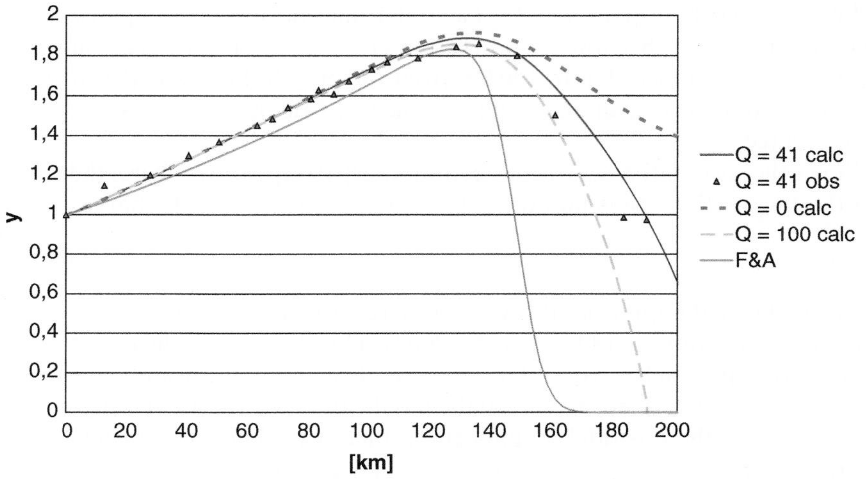

Figure 3.5 Tidal damping in the Schelde estuary on 21 June 1995 $(Q = 41\,\mathrm{m^3/s})$ showing the dimensionless tidal range $y = H/H(0)$ against distance. The drawn line through the observations (triangles) and the dotted lines have been computed by Horrevoets et al. (2004) for different river discharges. The lower curve has been computed with the method of Friedrichs and Aubrey (1994).

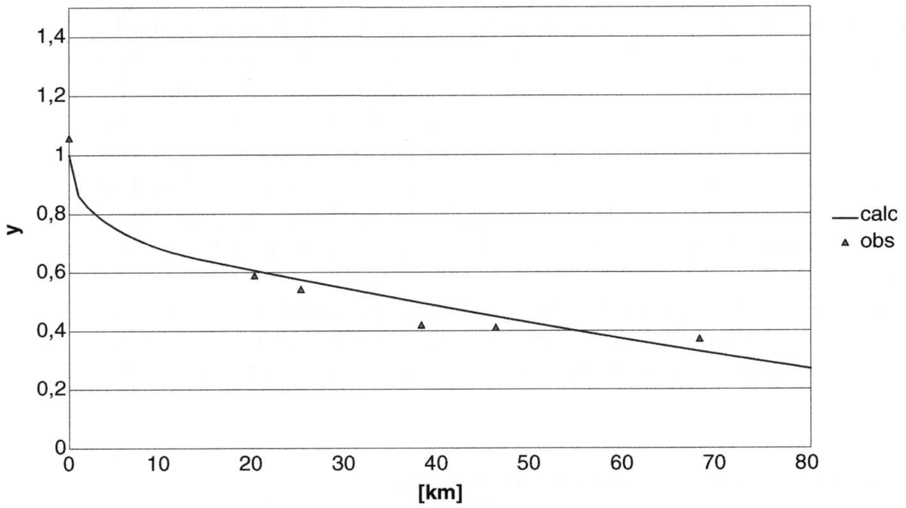

Figure 3.6 Tidal damping in the Incomati estuary on 30 July 1980 ($Q = 4\,m^3/s$) showing the dimensionless tidal range $y = H/H(0)$ against distance.

effect of the river discharge becomes important at the upstream tail of the estuary, resulting in a diversion between observed and calculated damping. The line for $Q_f = 0$ is the line obtained by Equation 3.23. The other lines have been computed with the equations derived in Section 3.3 where the effect of river discharge on tidal damping will be determined. Up to 150 km from the mouth, however, the fit is remarkably good. For comparison, the result obtained by Friedrichs and Aubrey (1994) is also presented.

Figure 3.6 shows the Incomati case which is continuously damped. Also here, the fit is good. In the Incomati, there is no noticeable influence from river discharge.

3.1.4 Conclusions

In perturbation analysis, damping is always considered to be exponential (e.g. Jay, 1991; Friedrichs and Aubrey, 1994; Prandle, 2003). This, however, is incorrect and has lead to inaccurate results, particularly in case of tidal amplification.

In alluvial estuaries, there is no exponential amplification of the tidal range along the estuary axis. If there is tidal amplification, then it is almost completely linear. There appears to be a negative feedback in tidal propagation preventing the occurrence of exponential amplification of the tidal range. In this feedback, the phase lag ε between HW and HWS is instrumental. Tidal amplification is dominant in estuaries with strong convergence. In short estuaries, this can lead to the development of a near standing wave where the wave type number ($N_E = \sin \varepsilon$) approaches zero. It can be seen from Equation 3.21 that this could lead to maximum amplification (the shortest amplification length scale $\beta = b$). However, if the tidal wave number approaches zero, then α, the tidal Froude number, approaches zero as well. This in turn makes the tidal amplification predominantly linear (see Equation 3.25) and even reduces tidal amplification to zero, (see Equation 3.23). This was indeed observed by Wright et al. (1973) who found almost no amplification and a near standing wave in the Ord estuary in Australia.

In estuaries where there is tidal damping, the process is also predominantly linear, particularly in the reach close to the estuary mouth where y is close to unity. As one moves further inland, the tidal range decreases and gradually the exponential term becomes dominant as y approaches zero. In estuaries with tidal damping, the process is primarily exponential upstream of the point where $x = -\beta/\alpha$. Downstream of that point the damping is primarily linear.

The fact that the wave type number is a crucial parameter in tidal propagation, makes it a prime variable to be observed during hydrometric surveys in estuaries. However, the phase lag between HW and HWS is seldom observed systematically. This requires an adjustment of measuring protocols.

Finally, fitting Equation 3.25 to observations of tidal propagation is an indirect way of measuring roughness and average estuary depth. Being aware that in estuaries C ranges between 40 and 70, this method can provide an estimate of average estuary depth which is otherwise difficult to obtain.

3.2 TIDAL WAVE PROPAGATION

Observations in estuaries indicate that an amplified tidal wave moves considerably faster than is indicated by the classical equation for wave propagation. Similarly, the celerity of propagation is lower if the tidal wave is damped. This phenomenon is clearly observed in the Schelde estuary (located in the Netherlands and Belgium) and in the Incomati estuary in Mozambique. In the Incomati, the tidal wave is damped throughout and the celerity of the wave is low as expected. In the Schelde, the tidal range increases from the estuary mouth as far as the city of Antwerp, after which it decreases until it reaches Gent. In harmony with the amplification and the damping, the tidal wave moves faster than expected in the downstream reach and slower in the upstream reach. This section presents a new analytical expression for the celerity of the tidal wave that takes into account the effect of tidal damping as an expansion of the classical equation for tidal wave propagation. The equation is successfully applied to observations in the Schelde and the Incomati. The new equation, developed by Savenije and Veling (2005), is a completely analytical solution of the St. Venant equations and appears to perform better than methods developed by earlier authors.

3.2.1 The relation between tidal damping and wave celerity

The classical formula for wave propagation is widely used to describe the propagation of a tidal wave in estuaries. This equation has been derived for the propagation of a small amplitude gravity wave in a channel of constant cross-section with no friction or bottom slope:

$$c_0^2 = \frac{1}{r_S} gh \tag{3.27}$$

where c_0 is the classical celerity of the wave (m/s), and all other parameters have been defined earlier. Under these conditions, the water level and flow are in phase.

The fact that this equation is so widely used in estuaries is surprising since the conditions for its derivation (constant cross section and no friction) do not apply in alluvial estuaries where the cross section varies exponentially along the estuary axis and friction is clearly not negligible. It will be demonstrated that the classical wave equation also describes the propagation of a tidal wave in a converging channel, with a phase lag ε between high water (HW) and high water slack (HWS), as long as it does not gain or lose amplitude as it travels upstream and the energy per unit width that is present in the wave is constant. This is the case when the energy gain from convergence of the banks, as the wave travels upstream, is compensated by the energy lost by friction. In this situation, one speaks of an 'ideal estuary' (Pillsbury, 1939), namely an estuary of constant depth, an exponentially varying width, a constant wave celerity and a constant phase lag between water level and velocity (see Section 2.2). The exponential width variation has been used widely to derive analytical equations for tidal wave propagation (see e.g. Hunt, 1964; Harleman, 1966; Jay, 1991, Savenije, 1992b; Friedrichs and Aubrey, 1994; Lanzoni and Seminara, 1998, 2002; Savenije and Veling, 2005). Although in an ideal estuary there is no tidal damping or amplification, in real estuaries there generally is, albeit modest. The length scale of tidal damping is generally large in relation to the length scale of bank convergence (see e.g. Friedrichs and Aubrey, 1994, Savenije, 1992b).

There appears to be a close relation between tidal damping (or amplification) and wave celerity. Tidal damping and tidal wave celerity both react to the imbalance between convergence and friction. In estuaries where tidal damping or amplification is apparent (e.g. the Thames, the Schelde, or the Incomati), one can observe a prominent deviation from the classical wave celerity c_0. If the wave is amplified, such as in the lower parts of the Thames and the Schelde, then the wave moves considerably faster than the celerity computed by Equation 3.27. When the wave is damped, as is the case in the Incomati and in upper parts of the Thames and Schelde, then the wave travels considerably slower.

Observations of tidal wave celerity under tidal damping and amplification are presented in Figure 3.7 for the Schelde in The Netherlands and in Figure 3.8 for the Incomati in Mozambique. These observations are combined with a drawn line representing the classical equation for wave propagation of Equation 3.27. Figure 3.7 shows observations at high water (HW) and low water (LW) in the Schelde and Figure 3.8 observations at high water slack (HWS) and low water

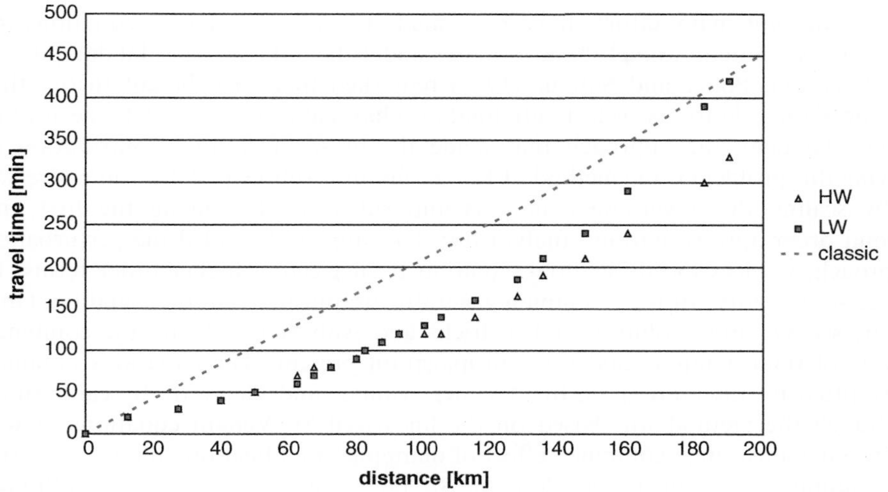

Figure 3.7 Observed propagation of the tidal wave at HW and LW on 21 June 1995 in the Schelde, compared to the propagation computed with the classical equation (Equation 3.27).

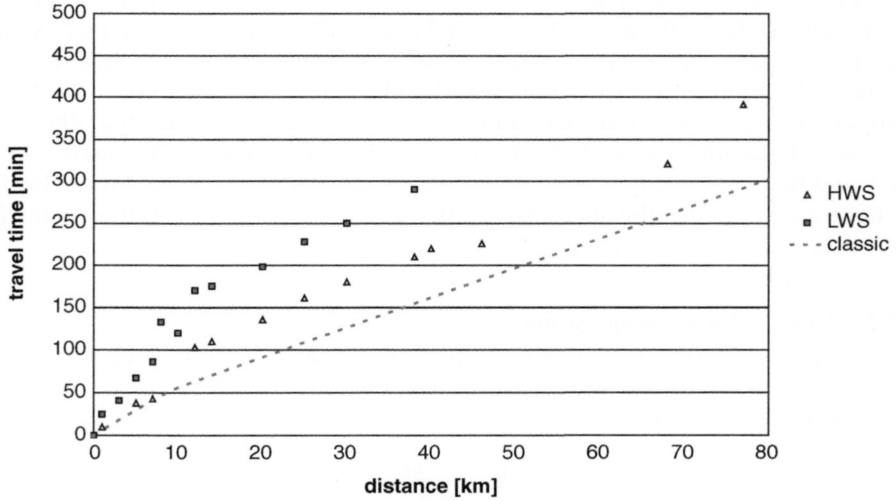

Figure 3.8 Observed propagation of the tidal wave at HWS and LWS on 23 June 1993 in the Incomati, compared to the propagation computed with the classical equation (Equation 3.27).

slack (LWS) in the Incomati. It can be seen clearly that in an amplified estuary (the lower Schelde) the travel time of the wave is much shorter than the travel time computed by Equation 3.27 (about half) and that the travel time is substantially larger in a damped estuary (the upper Schelde and the Incomati).

To date, not many efforts have been made to derive analytical equations for wave propagation in funnel shape estuaries under the influence of tidal damping or amplification. Ponce and Simons (1977) remarked that: 'a coherent theory that accounts for celerity as well as attenuation characteristics has yet to be formulated.' To date, the approach that made the most substantial contribution to solving this problem is the method of Friedrichs and Aubrey (1994), who addressed it by scaling the governing equations and subsequently solving the first and second order approximations analytically. This approach, called the perturbation approach, reduces the differential equation by neglecting higher order terms. In doing so, the effect of tidal damping generally disappears from the equation. Only in the second order solution of Friedrichs and Aubrey (1994) are the combined effects of tidal damping and wave propagation present. The equations obtained were fitted to observations in the Thames, Tamar, and Delaware by calibration. Although the method was based on the linearized St. Venant equations, it was able to demonstrate a combined effect of damping/amplification and wave celerity by retaining an exponentially decaying or increasing tidal amplitude (although we know now that tidal damping or amplification is predominantly linear).

In this section, the method of characteristics (see e.g. Dronkers, 1964; Whitham, 1974) is used to derive an analytical equation for the combined effect of damping and wave celerity. This equation allows the computation of the celerity of propagation under damped or amplified conditions. The equation obtained is not very complicated and is an extension of the classical formula in which, besides the rate of amplification (or damping), the phase lag ε between high water and high water slack is essential. Conditions for these derivations are that the tidal amplitude-to-depth ratio and the Froude number are considerably smaller than unity and that the velocity of river discharge is small compared to the tidal velocity. In the lower part of alluvial estuaries, this situation is the rule rather than the exception.

3.2.2 Theory of wave propagation

Rewriting the St. Venant equations (Equations 2.2 and 2.23), making use of the geometric relations (Equations 2.7–2.9) and the exponential width convergence (Equation 2.39), gives us:

$$r_S\frac{\partial h}{\partial t}+h\frac{\partial U}{\partial x}+U\frac{\partial h}{\partial x}-\frac{Uh}{b}=0 \tag{3.28}$$

$$\frac{\partial U}{\partial t}+U\frac{\partial U}{\partial x}+g\frac{\partial h}{\partial x}+gI_b-gI_r+f\frac{U|U|}{h}=0 \tag{3.29}$$

where all the parameters have been defined earlier.

The main unknown parameters to be determined in the analysis, as functions of time and space, are the water depth h and the water velocity U. Both are nsidered periodic functions Φ and Ψ with a tidal period T and an angular locity $\omega = 2\pi/T$.

Like the assumptions made on the character of the tidal wave in Equations 48–2.50, the following periodic functions can be applied for velocity and depth:

$$U = \upsilon\Phi(\xi - \varepsilon) \approx -\upsilon \sin(\xi - \varepsilon) \tag{3.30}$$

$$h = \eta\Psi(\xi) + \bar{h} \approx \eta \cos(\xi) + \bar{h} \tag{3.31}$$

$$\xi = \omega\left(t - \frac{x}{c}\right) + \xi_0 \tag{3.32}$$

ıere: $\upsilon(x)$ is the amplitude of the tidal velocity, $\eta(x)$ is the tidal amplitude $= H/2$), $\bar{h}$ is the tidal average depth, ξ is the dimensionless argument of the riodic function, c is the celerity of the tidal wave, and ε is the well-known ase lag between high water (HW) and high water slack (HWS), or between w water (LW) and low water slack (LWS). This phase lag can be assumed to constant along the estuary axis if the tidal amplitude-to-depth ratio and river charge is small compared to the tidal flow. The functions Φ and Ψ are harmonics nilar to sinusoids, but do not have to be exactly equal to them. They may r example contain higher order harmonics.

For the Equations 3.30–3.32 to be correct, the following conditions should be met:

The tidal velocity U in Equation 3.30 should not be influenced by the river discharge Q_f. Hence, $\upsilon \gg Q_f/A$.
The amplitude of the tidal water level variation is smaller than the depth of flow. Hence, $\eta < h$.
The Froude number is small: $\upsilon \ll c$.
The phase lag ε is constant along the estuary.
The wave celerity is constant along the estuary, or at least along a certain reach of the estuary, or $\partial c/\partial x = 0$.
The scaled tidal wave (Ψ) is not deformed as it travels upstream.

ıe first assumption is very common in tidal hydraulics and quite acceptable in the wnstream part of alluvial estuaries. The second is only correct in deep estuaries. a shallow estuary such as the Incomati, this situation does not apply near the uary mouth, although it does further upstream. Assumption 3 is linked to the cond assumption; we saw this in Equation 2.92. The Froude number is smaller ın the amplitude-to-depth ratio by a factor $r_S \sin(\varepsilon)$. Assumption 4 corresponds the theory of the ideal estuary, which in alluvial estuaries is a good approx- ation of reality. The fifth assumption is crucial for the derivations made. It also in agreement with the theory of the ideal estuary. Its validity shall be ecked in Section 3.2.3 for the estuaries studied. Assumption 6, finally, is the

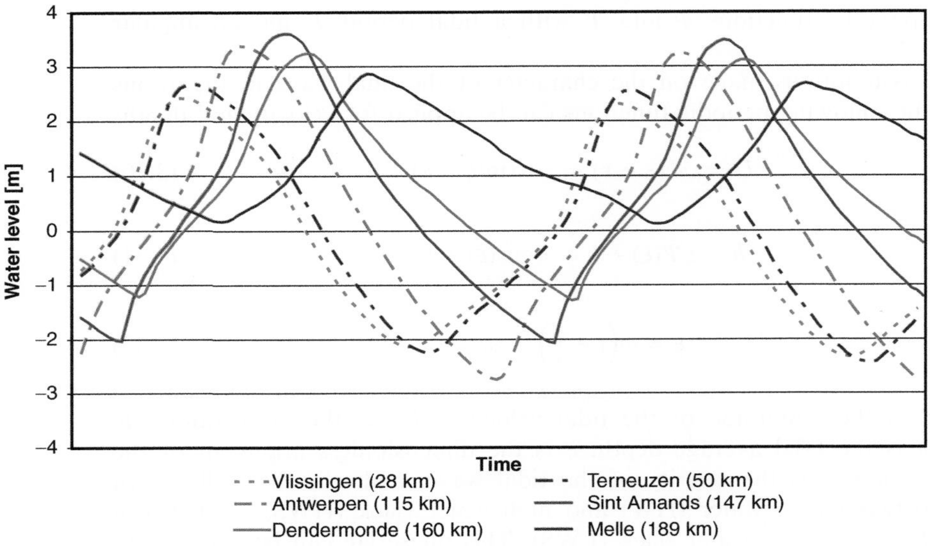

Figure 3.9 Observed tidal waves at different points along the Schelde estuary on 23 June 2001.

methodological assumption. This assumption, which implies that higher order effects are negligible, is less restrictive than the assumption made for the derivation of the classical equation. Assumption 6 is acceptable as long as assumptions 1–3 are valid, and therefore it is not really an additional assumption.

In Figure 3.9, we see the different stages of the tidal wave as it travels up the Schelde estuary. We see that the wave is first amplified and subsequently damped. We also see that it is gradually deformed by secondary effects, particularly the fact that friction is higher during LW and less during HW. Overall, however, the scaled wave, divided by its tidal range, retains its shape represented by the periodic function Ψ.

The following partial derivatives can be derived for the argument of the harmonic functions:

$$\frac{\partial \xi}{\partial t} = \omega \tag{3.33}$$

$$\frac{\partial \xi}{\partial x} = -\frac{\omega}{c} \tag{3.34}$$

The average water depth gradient can be found by averaging Equation 3.29 over time under the assumption that the residual effect of the friction term is small.

This is only correct under the strict assumptions of a small tidal range to depth ratio and negligible river flow:

$$\frac{\partial \bar{h}}{\partial x} = I_{\mathrm{r}} - I_{\mathrm{b}} \tag{3.35}$$

Subsequently, the following partial derivatives of velocity and depth can be written as:

$$\frac{\partial U}{\partial t} = \upsilon \Phi' \frac{\partial \xi}{\partial t} = \upsilon \omega \Phi' \tag{3.36}$$

$$\frac{\partial U}{\partial x} = \upsilon \Phi' \frac{\partial \xi}{\partial x} + \Phi \frac{\partial \upsilon}{\partial x} = -\frac{\upsilon}{c} \omega \Phi' + \upsilon \delta_{\mathrm{U}} \Phi \tag{3.37}$$

$$\frac{\partial h}{\partial t} = \eta \Psi' \frac{\partial \xi}{\partial t} = \eta \omega \Psi' \tag{3.38}$$

$$\frac{\partial h}{\partial x} = \eta \Psi' \frac{\partial \xi}{\partial x} + \Psi \frac{\partial \eta}{\partial x} + \frac{\partial \bar{h}}{\partial x} = -\eta \frac{\omega}{c} \Psi' + \eta \delta_{\mathrm{H}} \Psi + I_{\mathrm{r}} - I_{\mathrm{b}} \tag{3.39}$$

where δ_{U} and δ_{H} are the tidal damping rate for the tidal velocity amplitude and the tidal range, which were defined earlier in Equations 2.58 and 2.72, respectively. Note that for these derivatives, it is not necessary to assume that the periodic functions are simple harmonics.

Essential for further derivation is the assumption that the wave may be amplified or damped, but that the scaled tidal wave (Ψ) is not deformed, as long as the observer travels along a characteristic at the wave celerity (Ψ is constant with time if ξ is constant with time, implying that $\mathrm{d}x/\mathrm{d}t = c$). As we can see from Figure 3.9, showing the propagation of tidal waves in the Schelde estuary, the tide has a higher order M_4 component resulting from the non-linear resistance term. The shape of the scaled tidal wave, however, remains similar as the wave travels upstream.

Hence, for an observer travelling at the celerity of the tidal wave:

$$\frac{\mathrm{d}\Psi(x,t)}{\mathrm{d}t} = \frac{\partial \Psi}{\partial t} + c \frac{\partial \Psi}{\partial x} = \frac{\partial \Psi}{\partial \xi} \frac{\partial \xi}{\partial t} + c \frac{\partial \Psi}{\partial \xi} \frac{\partial \xi}{\partial x} = \Psi' \left(\frac{\partial \xi}{\partial t} + c \frac{\partial \xi}{\partial x} \right) = 0 \tag{3.40}$$

The crux of the method is that application of Equation 3.40 to the combined St. Venant equations implies that the sum of all terms containing Ψ' should be zero. These terms should be found in a linear combination of the two equations. Thus, the equation of continuity (Equation 3.28) is multiplied by a constant factor

Table 3.1 Representation of the terms of the St. Venant equations as functions of Ψ', Φ', Ψ, and Φ

Term	Ψ'	Φ'	Ψ	Φ	Constant
$r_S \dfrac{\partial h}{\partial t}$	$r_S \eta\omega$				
$-\dfrac{Uh}{b}$				$-\dfrac{\upsilon h}{b}$	
$h\dfrac{\partial U}{\partial x}$		$-h\dfrac{\upsilon\omega}{c}$		$+h\upsilon\delta_U$	
$U\dfrac{\partial h}{\partial x}$	$-U\dfrac{\omega}{c}\eta$		$U\eta\delta_H$	$-\upsilon(I_b - I_r)$	
$\dfrac{\partial U}{\partial t}$		$\upsilon\omega$			
$U\dfrac{\partial U}{\partial x}$		$-U\dfrac{\upsilon\omega}{c}$		$+U\upsilon\delta_U$	
$g\dfrac{\partial h}{\partial x}$	$-g\dfrac{\omega}{c}\eta$		$g\eta\delta_H$		$-g(I_b - I_r)$
$g(I_b - I_r)$					$g(I_b - I_r)$
$\dfrac{fU\lvert U\rvert}{h}$				$\dfrac{f\upsilon\lvert U\rvert}{h}$	

m and added to the equation of motion (Equation 3.29), after which the sum of all terms containing Ψ' are equated to zero. Such a method is more often used to determine the celerity of propagation, for example by Sobey (2001), who applied it to a channel of constant cross section and disregarded the effect of tidal damping. Our approach generates more terms.

To enhance insight into the terms, the St. Venant equations are represented in Table 3.1 in a format where each column lists the coefficients of the equations belonging to the variables Ψ', Φ', Ψ, and Φ for each term of the equations. In this Table, there are a number of cells that contain non-linear terms. These terms could also have been placed in another column. The reason why certain columns are chosen is to provide a logical overview. The position of these terms is not affecting further analysis. What is essential for the following analysis is to see which terms belong to Ψ'.

The combined equation can be split into two parts: the equation where the sum of the coefficients of the terms containing Ψ' is zero (corresponding to Equation 3.40), and the equation where the sum of the remaining terms is zero. The first equation yields:

$$m\left(r_S\eta\omega - U\frac{\omega}{c}\eta\right) - g\frac{\omega}{c}\eta = 0$$

or:

$$m = \frac{g}{(cr_S - U)} \tag{3.41}$$

where m is the multiplication factor for the equation of continuity.

The second equation reads:

$$\begin{aligned} &m\left\{-h\frac{\upsilon\omega}{c}\Phi' - \upsilon h\left(\frac{1}{b} - \delta_U\right)\Phi + \upsilon\Phi(\eta\delta_H\Psi + I_r - I_b)\right\} \\ &\quad + \upsilon\omega\left(1 - \frac{U}{c}\right)\Phi' + g\eta\delta_H\Psi + \left(\frac{f\upsilon|U|}{h} + U\upsilon\delta_U\right)\Phi = 0 \end{aligned} \tag{3.42}$$

Substitution of m and rearrangement yields:

$$(cr_S - U)(c - U) = gh\frac{1 + C + S + D_2 + D_3}{1 - (R + D_1 + D_4)(c/(c - U))} = gh\vartheta \tag{3.43}$$

where ϑ is the damping factor, which is a function of Φ and Ψ, consisting of the following terms:

$$C = \frac{\Phi c}{\Phi'\omega}\frac{1}{b}$$

$$S = \frac{\Phi c}{\Phi' h\omega}(I_b - I_r)$$

$$R = -\frac{\Phi}{\Phi'\omega}\frac{f|U|}{h}$$

$$D_1 = -\frac{g}{\Phi'\upsilon\omega}\eta\Psi\delta_H$$

$$D_2 = -\frac{\Phi c}{\Phi'\omega}\delta_U$$

$$D_3 = -\frac{\Phi c}{\Phi' h\omega}\eta\Psi\delta_{\mathrm{H}}$$

$$D_4 = -\frac{\Phi}{\Phi'\omega}\upsilon\Phi\delta_{\mathrm{U}}$$

These terms are all functions of U and h, and hence of time and space. Here C is the term that determines the acceleration of the wave due to the convergence of the banks. S determines the influence of the bottom slope. R determines the deceleration due to friction and the D_i terms ($i=1, \ldots, 4$) contain the effect of tidal amplification or damping. The terms are defined in such a way that if they are positive, they cause the wave to move faster. If they are negative, they slow down the propagation of the wave. Because in alluvial estuaries the Froude number ($\boldsymbol{F}=U/c$) is much smaller than unity: $|D_4| \ll |D_2|$ and $|\mathrm{D}_3| \ll |D_1|$. As a result, D_4 and D_3 can generally be disregarded, but we retain them here.

Airy's Equation

Besides that ϑ is a function of U and h, Equation 3.43 contains the flow velocity and the depth explicitly. This dependency on U and h can be simplified using an adjusted Airy equation. Airy (1845), quoted by Lamb (1932; art.175), presented the following equation for a frictionless undamped progressive wave ($\sin\varepsilon=\pi/2$) in a prismatic channel with no bottom slope:

$$c = \bar{c}_0\left(1 + \frac{3}{2}\frac{(h-\bar{h})}{\bar{h}}\right) \tag{3.44}$$

where $\bar{c}_0$ is the classical celerity of the tidal wave at mean depth. In our case, where there is friction, tidal damping, and a strong topography, this equation is different. For a small Froude number, Equation 3.43 can be modified as:

$$(c-U)^2 = \bar{c}_0^2\left(1 + \frac{h-\bar{h}}{\bar{h}}\right)\vartheta \tag{3.45}$$

Let us consider the situation at HW. At HW: $h-\bar{h}=\eta$ and $U=\upsilon\sin\varepsilon$. Making use of the Scaling equation, Equation 2.92, this leads to:

$$c_{\mathrm{HW}} = \bar{c}_0\left(1 + \frac{\eta}{\bar{h}}\right)^{0.5}\vartheta_{\mathrm{HW}}^{0.5} + \bar{c}_0 r_{\mathrm{S}}\frac{\eta}{\bar{h}}\sin^2\varepsilon \tag{3.46}$$

The root in the first term can be replaced by the first terms of a Taylor series expansion, if $\eta/h < 1$. Hence:

$$c_{HW} = \bar{c}_0\left(1+\frac{\eta}{2h}\right)\vartheta_{HW}^{0.5} + \bar{c}_0 r_S \frac{\eta}{h}\sin^2\varepsilon = \bar{c}_0\left(1+\frac{\eta}{h}\left(\frac{1}{2}+r_S\frac{\sin^2\varepsilon}{\vartheta_{HW}^{0.5}}\right)\right)\vartheta_{HW}^{0.5} \quad (3.47)$$

Now it can be seen that for $r_S = 1$, $\vartheta = 1$ and $\varepsilon = \pi/2$ (the case of an undamped progressive wave in a prismatic channel) this is the same as Airy's equation. In alluvial estuaries, however, the value of $\sin^2\varepsilon$ is O(0.1). With ϑ and r_S being close to unity, this implies that in alluvial estuaries the effect of the wave amplitude on the wave celerity is less strong as Airy's equation suggests. The general equation for the effect of depth and velocity variation on wave propagation can be derived similarly as:

$$c = \bar{c}_0\left(1-\frac{\eta}{2h}\Psi\right)\vartheta^{0.5} - \bar{c}_0 r_S \frac{\eta}{h}\Phi\sin\varepsilon = \bar{c}_0\left(1-\frac{\eta}{h}\left(\frac{\Psi}{2}+r_S\frac{\Phi\sin\varepsilon}{\vartheta^{0.5}}\right)\right)\vartheta^{0.5} \quad (3.48)$$

For a small amplitude-to-depth ratio, the direct effect of the water level fluctuation on the wave celerity is small. But there is a stronger effect through ϑ that we have not yet explored.

A solution for HWS and LWS

The values of C, S, R, and D_i vary during the tidal cycle. In Table 3.2, the values of these terms are presented for special moments during the tidal cycle: HW, LW, HWS, LWS, the tidal average situation (TA), and at maximum flow (MAX) during ebb and flood respectively. In this table, it is assumed that the functions Ψ and Φ are sinusoidal.

If the tidal wave progresses as a non-deformed wave, any convenient moment during the tidal cycle can be selected to determine the wave celerity. It can be seen from Table 3.2 that it is very attractive to solve the equation for the moments of HWS and LWS where: $U=0$, $C=0$, $S=0$, $R=0$, $D_2=0$, $D_3=0$, and $D_4=0$. Hence, the equation for the celerity of the wave (at slack time) reads:

$$c^2 = \frac{1}{r_S} gh \frac{1}{1-D} = \frac{c_0^2}{\left(1-\frac{g}{\upsilon\omega}\frac{d\eta}{dx}\cos\varepsilon\right)} \quad (3.49)$$

where $D = D_1$ and h is the water depth at slack time. Since the water depth at HWS is larger than at LWS, the celerity of the wave is higher at HWS than at LWS. Hence, the assumption that the wave is non-deformed is only valid if the depth at HWS is not much different from the depth at LWS, or if $\eta\cos\varepsilon \ll h$. In that case, the average depth may be used.

Equation 3.49 is an expansion of the classical equation for wave propagation by a simple damping factor $1/(1-D)$. The equation obtained is surprisingly simple

Table 3.2 Values of terms determining tidal propagation for HW, HWS, LW, LWS, TA and MAX flow situations

	HW	LW	HWS	LWS	TA ebb	TA flood	MAX ebb	MAX flood
ξ	0	π	ε	$\pi+\varepsilon$	$\pi/2$	$3\pi/2$	$\pi/2+\varepsilon$	$3\pi/2+\varepsilon$
ϕ	$\sin\varepsilon$	$-\sin\varepsilon$	0	0	$-\cos\varepsilon$	$\cos\varepsilon$	-1	$+1$
Φ'	$-\cos\varepsilon$	$\cos\varepsilon$	-1	1	$-\sin\varepsilon$	$\sin\varepsilon$	0	0
Ψ	1	-1	$\cos\varepsilon$	$-\cos\varepsilon$	0	0	$-\sin\varepsilon$	$\sin\varepsilon$
C	$-\frac{c\tan\varepsilon}{\omega}\frac{1}{b}$	$-\frac{c\tan\varepsilon}{\omega}\frac{1}{b}$	0	0	$\frac{c}{\omega b\tan\varepsilon}$	$\frac{c}{\omega b\tan\varepsilon}$	$-\frac{c}{\omega b}$	$\frac{c}{\omega b}$
S	$-\frac{c\tan\varepsilon}{h_{\mathrm{HW}}\omega}(I_{\mathrm{b}}-I_{\mathrm{r}})$	$-\frac{c\tan\varepsilon}{h_{\mathrm{LW}}\omega}(I_{\mathrm{b}}-I_{\mathrm{r}})$	0	0	$\frac{c(I_{\mathrm{b}}-I_{\mathrm{r}})}{\bar{h}\omega\tan\varepsilon}$	$\frac{c(I_{\mathrm{b}}-I_{\mathrm{r}})}{\bar{h}\omega\tan\varepsilon}$	$-\frac{c(I_{\mathrm{b}}-I_{\mathrm{r}})}{\bar{h}\omega}$	$\frac{c(I_{\mathrm{b}}-I_{\mathrm{r}})}{\bar{h}\omega}$
R	$\frac{f\upsilon(\sin\varepsilon)^2}{h_{\mathrm{HW}}\omega\cos\varepsilon}$	$\frac{f\upsilon(\sin\varepsilon)^2}{h_{\mathrm{LW}}\omega\cos\varepsilon}$	0	0	$\frac{-f\upsilon\cos\varepsilon}{\bar{h}\omega}$	$\frac{-f\upsilon\cos\varepsilon}{\bar{h}\omega}$	$\frac{f\upsilon}{\bar{h}\omega}$	$-\frac{f\upsilon}{\bar{h}\omega}$
D_1	$\frac{g\eta\delta_{\mathrm{H}}}{\upsilon\omega}\frac{1}{\cos\varepsilon}$	$\frac{g\eta\delta_{\mathrm{H}}}{\upsilon\omega}\frac{1}{\cos\varepsilon}$	$\frac{g\eta\delta_{\mathrm{H}}}{\upsilon\omega}\cos\varepsilon$	$\frac{g\eta\delta_{\mathrm{H}}}{\upsilon\omega}\cos\varepsilon$	0	0	$\frac{g\eta\delta_{\mathrm{H}}}{\upsilon\omega}\sin\varepsilon$	$-\frac{g\eta\delta_{\mathrm{H}}}{\upsilon\omega}\sin\varepsilon$
D_2	$\frac{c\tan\varepsilon}{\omega}\delta_{\mathrm{U}}$	$\frac{c\tan\varepsilon}{\omega}\delta_{\mathrm{U}}$	0	0	$\frac{-c}{\omega\tan\varepsilon}\delta_{\mathrm{U}}$	$\frac{-c}{\omega\tan\varepsilon}\delta_{\mathrm{U}}$	$\frac{c}{\omega}\delta_{\mathrm{U}}$	$-\frac{c}{\omega}\delta_{\mathrm{U}}$
D_3	$\frac{c\tan\varepsilon}{h_{\mathrm{HW}}\omega}\eta\delta_{\mathrm{H}}$	$-\frac{c\tan\varepsilon}{h_{\mathrm{LW}}\omega}\eta\delta_{\mathrm{H}}$	0	0	0	0	$-\frac{c\sin\varepsilon}{\bar{h}\omega}\eta\delta_{\mathrm{H}}$	$-\frac{c\sin\varepsilon}{\bar{h}\omega}\eta\delta_{\mathrm{H}}$
D_4	$\frac{\upsilon\sin\varepsilon\tan\varepsilon}{\omega}\delta_{\mathrm{U}}$	$-\frac{\upsilon\sin\varepsilon\tan\varepsilon}{\omega}\delta_{\mathrm{U}}$	0	0	$\frac{\upsilon\cos\varepsilon}{\omega\tan\varepsilon}\delta_{\mathrm{U}}$	$-\frac{\upsilon\cos\varepsilon}{\omega\tan\varepsilon}\delta_{\mathrm{U}}$	$-\frac{\upsilon}{\omega}\delta_{\mathrm{U}}$	$-\frac{\upsilon}{\omega}\delta_{\mathrm{U}}$

d provides clear insight into the factors that influence tidal propagation. It can seen directly that the classical equation is obtained if convergence balances ction and there is no tidal damping or amplification. The wave is slowed wn under tidal damping ($\mathrm{d}H/\mathrm{d}x < 0$) and accelerated under tidal amplification $H/\mathrm{d}x > 0$). If we analyze the origin of the terms in the denominator, we see at '1' represents the acceleration term in the momentum balance equation. e second term represents the damping component of the depth gradient in the omentum balance equation.

What can we say of the wave celerity at TA? At TA, the depth is the same the flood and ebb tide. The asymmetry between TA flood and TA ebb is in the term and in the left hand side of Equation 3.43 where the velocity has a different n. Both cases of asymmetry are negligible if the Froude number is small. Dis- garding the D_4 term, one can demonstrate that the celerity at TA and HWS/LWS the same, making use of the Damping equation (Equation 3.17). Rearrangement Equation 3.17 yields the expression for tidal damping or amplification:

$$\frac{\mathrm{d}H}{\mathrm{d}x} = H\frac{\left(\frac{1}{b} - f'\frac{\upsilon \sin\varepsilon}{\bar{h}c}\right)}{\left(1 + \frac{g\eta}{c\upsilon\sin\varepsilon}\right)} = H\frac{\left(\frac{1}{b} - f'\frac{\upsilon \sin\varepsilon}{\bar{h}c}\right)}{\left(1 + \frac{1}{\alpha}\right)} \tag{3.50}$$

ich we can use to substitute in Equation 3.49, so as to obtain an implicit relation the wave celerity:

$$c^2 = \frac{1}{r_S} gh \frac{1}{1 - D}$$

th:

$$D = \frac{c\sin\varepsilon\cos\varepsilon}{\omega}\frac{\left(\frac{1}{b} - f'\frac{\upsilon\sin\varepsilon}{\bar{h}c}\right)}{(1+\alpha)} = \frac{\sin 2\varepsilon}{2(1+\alpha)}\left(\frac{c}{\omega b} - \frac{R'}{\omega}\right) \tag{3.51}$$

Equation 3.51, both the nominator and denominator of D have been multiplied α. As a result, the denominator $(1+\alpha)$ has a value close to unity. In the right nd side of Equation 3.51, the friction term of Jay (1991) is used, with $R' = f'\upsilon$ ε/h. For convergence, Jay (1991) used the parameter $\Delta l = c(b\omega)^{-1}$. It can be n clearly that $c = c_0$ when these two terms are equal. This can also be seen in , 1991, figure 3, where the lines of Equation 3.51 plot a similar pattern.

It is also interesting to look at what would happen if D equals unity. Since the in the denominator corresponds with the acceleration term, it implies that tical convergence occurs when convergence and acceleration cancel out. This is situation of critical convergence mentioned by Jay (1991). If we neglect the ect of friction, critical convergence occurs when b equals $c\sin(2\varepsilon)/(2\omega(1+\alpha))$. perturbation analysis, where the phase lag ε is disregarded, it is said that critical nvergence occurs when $b = c/2\omega$ (Jay, 1991). In real estuaries, this factor 2 lies tween 3 and 4 as a result of the phase lag ε. There is a very good feedback mechanism that prevents critical convergence from occurring. If the convergence length is very small, this will lead to the tidal wave being more strongly reflected, which reduces the value of ε. This in turn decreases D.

One can simplify Equation 3.51 by substitution of Equation 2.88.

$$D = \frac{\cos^2 \varepsilon}{(1 - \delta b)} \frac{\left(1 - f' b \frac{\upsilon \sin \varepsilon}{hc}\right)}{(1 + \alpha)} \tag{3.52}$$

Equation 3.52 shows that the damping term is determined by the phase lag ε and the balance between convection and friction. It can also be seen that D is always smaller than unity. (The only factor that can make D larger than unity is $(1 - \delta b)$, which in the Schelde can reach a minimum value of 0.82, while $\cos^2\varepsilon = 0.89$ and $(1 + \alpha) = 1.11$). Obviously, Equation 3.51 is not defined if D approaches unity.

3.2.3 Empirical verification in the Schelde and Incomati estuaries

The theory described in the previous section has been confronted with observations made in the Schelde and Incomati estuaries. Figures 3.10 and 3.11 show the application of the theory to wave celerity in the two estuaries. The new equation performs considerably better than the classical equation. In the Incomati, the correspondence is very good, with the exception of the part nearest to the mouth. In the

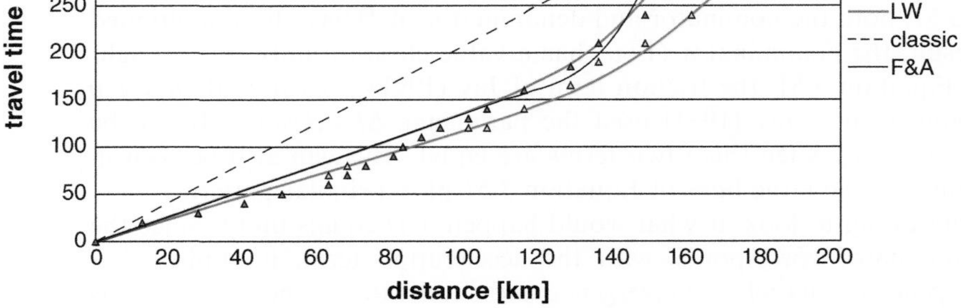

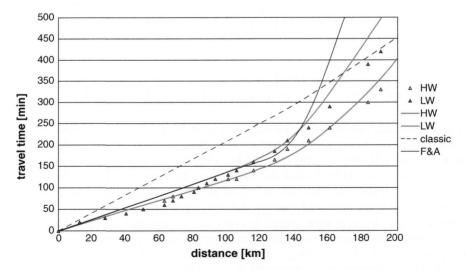

Figure 3.10 Propagation of the tidal wave in the Schelde estuary observed at high water (HW) and low water (LW) 21 June1995 (indicated by triangles). The thick drawn lines represent the computed wave propagation at HW and LW. The dotted line represents the classical wave propagation. The thin drawn line represents the wave propagation according to the method of Friedrichs and Aubrey (1994).

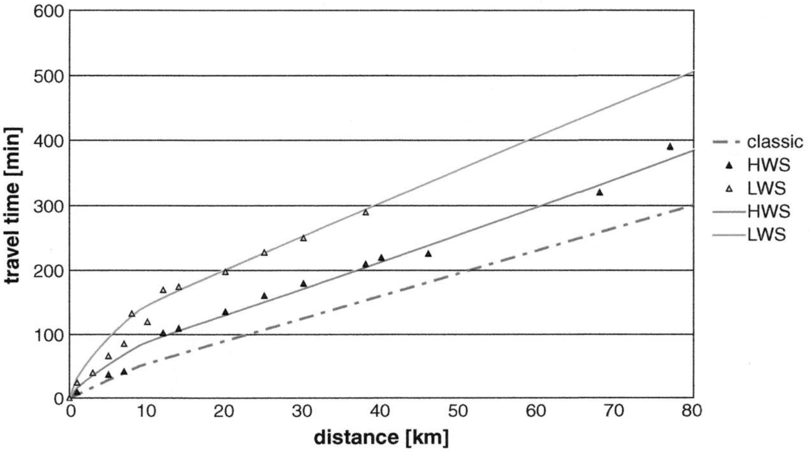

'igure 3.11 Propagation of the tidal wave in the Incomati estuary at high water slack HWS) and low water slack (LWS) observed on 23 June 1993 (indicated by triangles). The rawn lines represent the computed wave propagation at HWS and LWS. The dotted line epresents the classical wave propagation.

chelde the line for HW is very good, but the line for LW shows a deviation pstream from the point located 150 km from the mouth. The reasons for these eviations should be sought in the relatively high ratio of tidal amplitude-to-depth nd the effect of river discharge, of which the details are given by Savenije and 'eling (2005). For the sake of comparison, Figure 3.10 also shows the relation erived by Friedrichs and Aubrey (1994) for the tidal average situation.

Figure 3.12 shows the variation of the damping term D as a function of x in he Schelde estuary. The thick line is the value obtained by Equation 3.49 using bserved values of $\mathrm{d}H/\mathrm{d}x$. The line indicated by D''' is obtained from Equation .51. The reason why this line deviates from the previous line upstream of 140 km s because it does not take into account the effect of river discharge on tidal amping. This can be seen if we consider the thin line (indicated by D'), which vas obtained by using a slightly more sophisticated formula than Equation 3.50, eveloped by Horrevoets et al. (2004) taking into account the effect of river ischarge. The theory on the effect of river discharge on tidal damping is resented in Section 3.3. The line that takes river discharge into account fits the bserved data very well. Finally, the line indicated by D'' is the one using Equation .52 which makes use of Equation 2.88. The latter equation is only correct by pproximation and is more sensitive to high river flow and amplitude-to-depth atio. However, it can be seen that Equation 3.52 is a good approximation for he tide-dominated part of the Schelde estuary. In the Incomati, where the river ischarge is very small, we see that all curves except D'' are close (see Figure 3.13). 'he latter is not appropriate near the estuary mouth where the tidal amplitude-o-depth ratio is large.

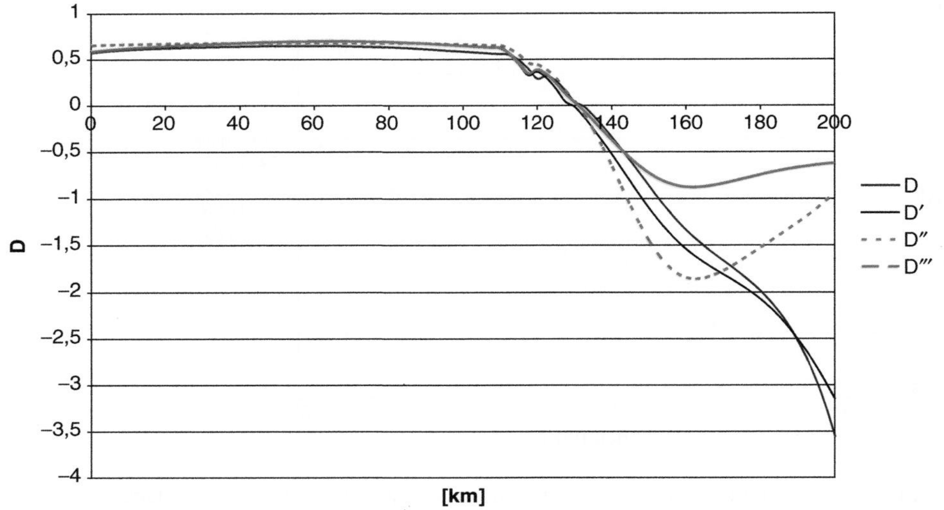

Figure 3.12 The variation of the damping term along the axis of the Schelde estuary. D indicates the damping term obtained by Equation 3.49, using observed values of dH/dx. D''' was obtained with Equation 3.51, D'' with Equation 3.52, and D' by using the equation of Horrevoets et al. (2004) that accounts for the effect of river discharge on tidal damping.

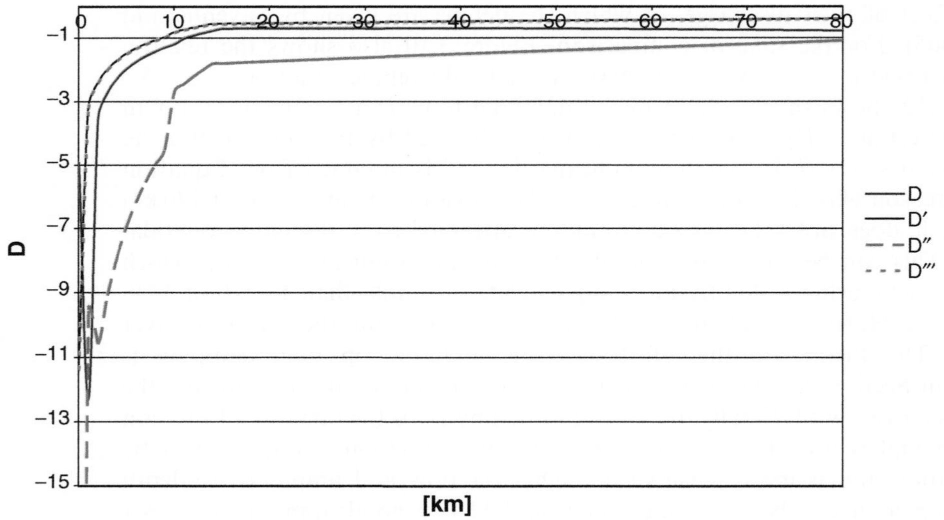

Figure 3.13 The variation of the damping term along the axis of the Incomati estuary. D indicates the damping term obtained by Equation 3.49, using observed values of dH/dx. D'' was obtained with Equation 3.51, D''' using Equation 3.52, and D' by using the equation of Horrevoets et al. (2004) that accounts for the effect of river discharge on tidal damping.

3.2.4 The wave celerity according to Mazure

Mazure (1937) presented an expression of the wave celerity based on the momentum equation. Equation 2.88 is based on the continuity equation and Equation 3.51 on the combination of both, but Mazure's equation uses purely the momentum equation, although—of course—both equations are used implicitly by assuming the periodic form of the solution. We can derive Mazure's solution from Equations 3.28–3.39. In addition we need an expression for the water level slope. The spatial derivative of the water level $Z = Z_b + h$ can be written as (dropping the subscript in δ_H):

$$\frac{\partial Z}{\partial x} = \eta \Psi' \frac{\partial \xi}{\partial x} + \Psi \frac{\partial \eta}{\partial x} + I_r = -\eta \frac{\omega}{c} \Psi' + \eta \delta \Psi + I_r \tag{3.53}$$

The equation of motion can be written as:

$$\frac{\partial U}{\partial t} + U \frac{\partial U}{\partial x} + g \frac{\partial Z}{\partial x} - g I_r + \frac{f U|U|}{h} = 0 \tag{3.54}$$

Substituting and writing the full derivative yields:

$$\frac{\mathrm{d}U}{\mathrm{d}t} + g \frac{\partial Z}{\partial x} - g I_r + \frac{f U|U|}{h} = 0 \tag{3.55}$$

Now let us consider the moment of maximum tidal velocity during the ebb current (MAX ebb in Table 3.2) this is the point where $U = -\upsilon$ and $\mathrm{d}U/\mathrm{d}t = 0$. Hence choosing $\xi_0 = 0$:

$$\xi = \pi/2 + \varepsilon,$$

$$\Psi = \cos(\pi/2 + \varepsilon) = -\sin \varepsilon,$$

$$\Psi' = -\sin(\pi/2 + \varepsilon) = -\cos \varepsilon$$

Substitution of these values into the conservation of momentum equation leads to:

$$g\left(-\eta \frac{\omega}{c} \Psi' + \eta \delta \Psi + I_r\right) - g I_r - \frac{f \upsilon^2}{h} = g\left(\eta \frac{\omega}{c} \cos \varepsilon - \eta \delta \sin \varepsilon\right) - \frac{f \upsilon^2}{h} = 0 \tag{3.56}$$

Hence:

$$c = \frac{\omega \eta g h \cos \varepsilon}{f \upsilon^2} \left(\frac{1}{1 + D_M}\right)$$

with:

$$D_M = \frac{g h \delta \eta \sin \varepsilon}{f \upsilon^2} \tag{3.57}$$

where it can be shown that D_M is a small number since $\delta\eta\sin\varepsilon$ is $O(10^{-6})$ which remains small compared to $f\,\boldsymbol{F}^2$, as long as damping is modest. Mazure (1937), quoting Canter Cremers (1921), derived the above equation by the same reasoning while disregarding the phase lag between HW and HWS, assuming that $\varepsilon = 0$, thereby obtaining:

$$c_{CC} = \frac{\omega\eta g h}{f\upsilon^2} \tag{3.58}$$

Since the term within parentheses is close to unity, particularly if damping is small, the mistake of these early researchers is essentially cos ε. Substitution of the 'Geometry-Tide' relation (Equation 2.90) yields:

$$c = \frac{\omega\eta\cos\varepsilon}{\upsilon}\frac{gh}{f\upsilon}\left(\frac{1}{1+D_M}\right) = \frac{(1-\delta b)}{r_S b}\frac{gh^2}{f\upsilon}\left(\frac{1}{1+D_M}\right) \tag{3.59}$$

The practical use of this formula is limited since the roughness f is seldom known beforehand. It may be used, however, to obtain an estimate of the channel roughness. Moreover, by making use of Equation 2.88 and assuming D_M is negligible, it can be demonstrated that this equation is essentially the same as the general equation for the wave celerity, Equation 3.51.

3.2.5. Conclusion

In the above derivations, a number of interesting new analytical equations have been formulated, which are summarized in Table 3.3. They constitute: the 'Phase Lag equation' (Equation 2.88), the 'Geometry-Tide relation' (Equation 2.90), the 'Scaling equation' (Equation 2.92), the 'Damping equation' (Equation 3.17) and the 'Celerity equation' (Equation 3.51). These equations are also compared to their 'classical' counterparts, which have been used widely in the past. Also Mazure's equation is presented, which can be demonstrated to be similar to Equation 3.51.

The main differences between the 'new' equations and their classical counterparts are: 1) the prominent presence of the phase lag ε, 2) the role attributed to tidal damping, and 3) the balance between convergence and friction. These three effects are often disregarded in classical analysis or treated in conflicting ways. In analyzing the effect of these factors, it is interesting to look at when a 'new' equation collapses into its classical counterpart. The Geometry-Tide relation equals the classical equation when damping is zero and $\varepsilon = 0$ (a standing wave). The Scaling equation, however, is the same as the classical counterpart when $\varepsilon = \pi/2$ (a progressive wave). Obviously, if one uses these two classical equations jointly in a certain situation, there is a serious lack of compatibility.

The classical counterpart of the Damping equation is 'Green's law' derived from the conservation of energy equation for a frictionless tidal wave. Jay (1991) demonstrated its limitations, but it is interesting to see the similarity with the

Table 3.3 Analytical equations describing tidal movement in alluvial estuaries.

Name	Equation number	Newly derived equation	'Classical' equation
Phase Lag equation	2.88	$\tan \varepsilon = \dfrac{\omega b}{c(1 - \delta b)} = \dfrac{b}{\lambda}\dfrac{2\pi}{(1 - \delta b)}$	
Geometry-Tide relation	2.90	$\dfrac{H}{E} = \dfrac{\eta \omega}{\upsilon} = \dfrac{\bar{h}}{r_S b}\dfrac{(1 - \delta b)}{\cos(\varepsilon)}$	$\dfrac{\eta \omega}{\upsilon} = \dfrac{\bar{h}}{r_S b}$
Scaling equation	2.92	$r_S \dfrac{\eta}{h} = \dfrac{\upsilon}{c}\dfrac{1}{\sin(\varepsilon)}$	$r_S \dfrac{\eta}{h} = \dfrac{\upsilon}{c}$
Damping equation	3.17	$\dfrac{\mathrm{d}H}{\mathrm{d}x}\left(1 + \dfrac{g\eta}{c\upsilon \sin \varepsilon}\right)$ $= H\left(\dfrac{1}{b} - f'\dfrac{\upsilon \sin \varepsilon}{\bar{h}c}\right)$	$\dfrac{\mathrm{d}H}{\mathrm{d}x} = H\dfrac{1}{2b}$
Celerity equation	3.51	$c^2 = \dfrac{1}{r_S}gh/\left[1 - \dfrac{\sin 2\varepsilon}{2(1 + \alpha)}\left(\dfrac{c}{\omega b} - \dfrac{R'}{\omega}\right)\right]$	$c^2 = \dfrac{1}{r_S}gh$
Mazure's equation	3.57	$c = \dfrac{\omega g h \eta \cos \varepsilon}{f \upsilon^2}$	$c = \dfrac{\omega g h \eta}{f \upsilon^2}$

general equation derived herein. Green's law considers convergence but assumes zero friction ($f' = 0$) and $\varepsilon = \pi/2$ (a progressive wave). If we assume constant depth and $r_S = 1$, Green's law follows from the Damping equation, by substitution of the Scaling equation.

The celerity equation equals its classical counterpart if friction and convergence are balanced so that damping is zero (the condition for the ideal estuary). Moreover, we see that the celerity equation collapses into the classical equation when $\sin 2\varepsilon = 0$, which is the case both for a standing and a progressive wave, irrespective of damping or amplification. For the standing wave, however, this conclusion is misleading, as Equation 3.52 shows us. For a purely standing wave, $\sin \varepsilon = 0$ and $\cos \varepsilon = 1$, resulting in D becoming close to unity and the celerity becoming infinitely large, which indeed happens.

Where the classical equations are contradictory in their assumptions (some assume a standing wave, some a progressive wave, some assume friction or convergence, others do not) the new equations in Table 3.3 are consistent and compatible. The new equations are general versions of their 'classical' counterparts, with the most important difference being the account they take of the phase lag between HW and HWS. This makes the Wave-type Number ($N_E = \sin \varepsilon$) a key parameter in tidal hydraulics. One can conclude that the new equations, while remaining simple, are a substantial improvement over the classical equations and that they provide new insight into the propagation of tidal waves under damped or amplified conditions. They are the result of an integrated theory for wave celerity and wave attenuation in which the phase lag between high water and high water slack plays a key role; the phase lag being a fundamental characteristic of an alluvial estuary, which is generally disregarded in perturbation analysis.

3.3 EFFECT OF RIVER DISCHARGE AND OTHER HIGHER ORDER EFFECTS ON TIDAL DAMPING

3.3.1 Which higher order effects are important

In the previous sections, a number of assumptions have been made regarding estuary hydraulics and shape. The most important were that:

1. the topography is that of an ideal estuary, with an exponentially varying width and no (or very modest) bottom slope;
2. the Froude number is small
3. the tidal amplitude-to-depth ratio is small
4. the river discharge is small compared to the tidal discharge
5. the phase lag is constant
6. the celerity is constant.

It appears that in the upstream part of the estuary the third and fourth assumptions become restrictive, particularly where it concerns the derivation of the Damping equation (as seen in Figure 3.5). The effect of river discharge is not always negligible and the tidal amplitude-to-depth ratio is not always very small. In shallow estuaries, or in estuaries that experience tidal amplification the tidal amplitude-to-depth ratio can approach unity. Both have a noticeable effect on the friction term: the discharge because the velocity during ebb is stronger than during flood and the depth because Chézy's roughness coefficient is depth-dependent and hence friction is stronger during ebb than during flood.

In the following, the effect of river discharge and a large amplitude-to-depth ratio on the derivation of the damping equation is explored. Subsequently a more accurate version is derived for a situation where river discharge can no longer be neglected. It will be shown that taking account of these effects primarily lead to a revised expression for the roughness term R'.

In the derivation of the equation for wave celerity, similar assumptions have been made. We shall see that river discharge does not influence the derivation of Equation 3.51 itself. If however we use the revised expression for the roughness term R' in Equation 3.51, we significantly improve the prediction of the wave celerity in the upstream part of an amplified estuary like the Schelde where the river discharge is not small compared to the tidal flow. The derivations presented in this section are based on the work of Horrevoets et al. (2004).

3.3.2 Incorporating river discharge into the derivation of the Celerity equation

The river discharge only affects the Celerity equation through U, resulting in a modification of Equation 3.30:

$$U = \upsilon\Phi(\xi - \varepsilon) - \frac{Q_\mathrm{f}}{A} \tag{3.60}$$

As a result, the velocity gradient and the friction term need to be modified. Modification of Equation 3.37 leads to:

$$\frac{\partial U}{\partial x} = -\frac{\upsilon}{c}\omega\Phi' + \upsilon\delta_\mathrm{U}\Phi - \frac{Q_\mathrm{f}}{aA} \tag{3.61}$$

Subsequently, the effect of the river discharge on the velocity gradient can be accounted for in the damping term D_4:

$$D_4 = -\frac{\Phi}{\Phi'\omega}\upsilon\Phi\left(\delta_\mathrm{U} + \frac{Q_\mathrm{f}}{A\upsilon a}\right) \tag{3.62}$$

Although D_4 is normally an order of magnitude smaller than D_2 because it is scaled by the Froude number, it may no longer be negligible if the river discharge is large compared to the tidal flow. However, at HWS and LWS (even with a significant upstream discharge) $D_4 \approx 0$ since $\Phi \approx 0$ and hence, there is no effect of the river discharge on this damping term.

The other possible effect of river discharge on wave propagation is through the friction term. The friction term needs to be adjusted as follows:

$$R = -\frac{1}{\Phi'\omega}\frac{f}{h\upsilon}\left(U - \frac{Q_\mathrm{f}}{A}\right)\left|U - \frac{Q_\mathrm{f}}{A}\right| \tag{3.63}$$

Slack is defined as the situation where the flow velocity is zero, which is the case when $U - Q_\mathrm{f}/A = 0$ and $R = 0$. So, there is no effect of river discharge on the friction term, but there is an influence through the phase lag. The river discharge causes a shift in the occurrence of slack with HWS occurring earlier and LWS occurring later. This can be easily seen if we shift the velocity curve downwards in Figure 2.5 over a distance Q_f/A. As the river discharge becomes more prominent, the phase lag near HW reduces and the phase lag at LW increases up to the point where they coincide. The phase lag near HW then equals ε-$\pi/2$ and the phase lag near LW equals ε-$\pi/2$. At this point, the phase lag near HW is negative and slack occurs earlier than HW (see Horrevoets et al., 2004). If the phase lag near HW reduces to zero, then the wave at HWS propagates at c_0, which is faster than the celerity of the damped tidal wave. Similarly, the wave near LW will appear to move slower than predicted as a result of the phase shift.

Hence, Equation 3.51 remains unaffected by river discharge, but the value of ε in Equation 3.51 may need to be adjusted for HWS and LWS. As a result, R' will be smaller during HWS and larger during LWS. This will lead to a faster wave during HWS and a slower wave during LWS. This effect can indeed be noticed in the Schelde as we shall see further on. Besides the effect of the phase lag on R', there is also the direct influence of the river discharge on R' which will be discussed in the next section.

3.3.3 Incorporating river discharge into the derivation of the Damping equation

In this section we use the Manning-Strickler equation (Equation 2.5) instead of the Chézy equation since the Chézy roughness coefficient C is depth-dependent, whereas Manning's K is not. Hence Equation 3.3 is modified into:

$$r_S \frac{cV}{gh}\frac{\mathrm{d}h}{\mathrm{d}x} - \frac{cV}{g}\left(\frac{1}{b} - \frac{1}{H}\frac{\mathrm{d}H}{\mathrm{d}x}\right) + \frac{\partial h}{\partial x} + I_b - I_r + \frac{V|V|}{K^2 h^{1.33}} = 0 \tag{3.64}$$

where K $(=1/n)$ is Manning's coefficient and V is the velocity of a water particle in a Lagrangean reference frame that moves with the water.

The same equations are used as in Section 3.1.2 for the depth and the depth gradients at HW and LW (Equations 3.4–3.9). However we modify the velocity to account for river discharge:

$$V_{\mathrm{HW}} = \upsilon \sin(\varepsilon) - \frac{Q_f}{A} \tag{3.65}$$

and similarly for LW:

$$V_{\mathrm{LW}} = -\upsilon \sin(\varepsilon) - \frac{Q_f}{A} \tag{3.66}$$

Here, the assumption is that the cross-sectional area of the stream at HW and LW is not much different. This simplification is acceptable as long as the tidal amplitude is small compared to the depth of flow. Combination of these equations leads to the following expression for HW:

$$\begin{aligned} r_S \frac{c_{\mathrm{HW}}(\upsilon \sin(\varepsilon) - Q_f/A)}{g h_{\mathrm{HW}}}\frac{\mathrm{d}h_{\mathrm{HW}}}{\mathrm{d}x} - \frac{c_{\mathrm{HW}}(\upsilon \sin(\varepsilon) - Q_f/A)}{g}\left(\frac{1}{b} - \frac{1}{H}\frac{\mathrm{d}H}{\mathrm{d}x}\right) \\ + \frac{\mathrm{d}h_{\mathrm{HW}}}{\mathrm{d}x} \pm \frac{(\upsilon \sin(\varepsilon) - Q_f/A)^2}{K^2 h_{\mathrm{HW}}^{1.33}} = -I_b + I_r \end{aligned} \tag{3.67}$$

The last term is the friction term and has a positive value if $\upsilon\sin(\varepsilon) > Q_f/A$ and a negative value if $\upsilon\sin(\varepsilon) < Q_f/A$.

Similarly for LW it follows that:

$$
\begin{aligned}
&-r_S\frac{c_{LW}(\upsilon\sin(\varepsilon)+Q_f/A)}{gh_{LW}}\frac{dh_{LW}}{dx}+\frac{c_{LW}(\upsilon\sin(\varepsilon)+Q_f/A)}{g}\left(\frac{1}{b}-\frac{1}{H}\frac{dH}{dx}\right)\\
&+\frac{dh_{LW}}{dx}-\frac{(\upsilon\sin(\varepsilon)+Q_f/A)^2}{K^2h_{LW}^{1.33}}=-I_b+I_r
\end{aligned}
\tag{3.68}
$$

To arrive at an expression for the tidal range H, these equations should be subtracted, leading to terms containing dH/dx or I. This is straightforward if the coefficients of the terms are the same, otherwise it leads to correction factors.

Interestingly, the ratio of c/h in the first term is similar for HW and LW since both the celerity and the depth are higher than average at HW and lower than average at LW. We shall therefore assume that these two coefficients are similar, so that the terms can be subtracted. As was observed in Section 3.1.2, the coefficients of the first term of Equations 3.67 and 3.68 are scaled a Froude number smaller than that of the third term, while the sum of dh/dx for HW and LW is equal to the residual slope I, which is small compared to dH/dx. Hence, the first terms in Equations 3.67 and 3.68 are small compared to the other terms.

In the second term of Equations 3.67 and 3.68, the celerity is different at HW and LW. As a result, a correction factor γ appears in the subsequent subtraction of these terms. In the last term (the resistance term), the water depth is different at HW and at LW. This also leads to a correction factor (f'').

Subsequently, Equations 3.67 and 3.68 are subtracted and combined with Equations 3.4 and 3.7. Details on the subtraction of individual terms are presented in Horrevoets et al. (2004). Subtraction yields the following expressions:

for *zone I*, where $\upsilon\sin(\varepsilon) > Q_f/A$:

$$
\begin{aligned}
&\frac{dH}{dx}\left(\frac{g}{2c\upsilon\sin(\varepsilon)}-\frac{Q_f}{A\upsilon\sin(\varepsilon)2\bar{h}}+\frac{\gamma}{H}\right)\\
&=\frac{\gamma}{b}-\frac{f''\upsilon\sin(\varepsilon)}{\bar{h}^{1.33}c}\left[1+\frac{1.33H}{\bar{h}}\frac{Q_f}{A\upsilon\sin(\varepsilon)}+\frac{Q_r^2}{A^2\upsilon^2\sin^2(\varepsilon)}\right]-\frac{I}{\bar{h}}
\end{aligned}
\tag{3.69}
$$

for *zone II*, where $\upsilon\sin(\varepsilon) < Q_f/A$:

$$
\begin{aligned}
&\frac{dH}{dx}\left(\frac{g}{2c\upsilon\sin(\varepsilon)}-\frac{Q_f}{A\upsilon\sin(\varepsilon)2\bar{h}}+\frac{\gamma}{H}\right)\\
&=\frac{\gamma}{b}-\frac{f''\upsilon\sin(\varepsilon)}{\bar{h}^{1.33}c}\left[\frac{1.33H}{2\bar{h}}+\frac{2Q_f}{A\upsilon\sin(\varepsilon)}+\frac{1.33H}{2\bar{h}}\frac{Q_r^2}{A^2\upsilon^2\sin^2(\varepsilon)}\right]-\frac{I}{\bar{h}}
\end{aligned}
\tag{3.70}
$$

where $\bar{h}$ is the tidal average depth, I is the slope of the average water level, γ is a correction factor for wave celerity and f'' is the adjusted friction factor (similar to Equation 3.15):

$$f'' = \frac{g}{K^2}\left(1 - \left(\frac{1.33\eta}{\bar{h}}\right)^2\right)^{-1} \tag{3.71}$$

$$\gamma = 1 - \frac{(c_{\mathrm{HW}} - c)}{c}\frac{Q_{\mathrm{f}}}{A\upsilon\sin(\varepsilon)} \approx 1 - \left(\sqrt{1+\frac{\eta}{\bar{h}}} - 1\right)\frac{Q_{\mathrm{f}}}{A\upsilon\sin(\varepsilon)} \tag{3.72}$$

The friction factor f'' compensates for the fact that friction is larger at LW than at HW and is always larger than unity. It enhances the effect of friction. If $\eta/h \ll 1$, $f'' \approx 1$. In the upper reaches of an amplified estuary however, the friction factor can become very important as η/h approaches unity (e.g. if $\eta/h = 0.5$, $f'' = 1.8$). The coefficient 1.33 in these equations follows from a Taylor series expansion of $(h+\eta)^{1.33} \approx h^{1.33}(1+1.33\ \eta/h)$, if $\eta < h$. Due to the factor 1.33, Equation 3.71 only makes sense as long as $\eta/h < 0.7$.

The correction factor γ compensates for the difference in wave celerity at HW and LW in the second term of Equations 3.67 and 3.68. It has a value smaller than unity, but is close to unity as long as $\eta/h \ll 1$ and $Q_{\mathrm{f}}/(A\ \upsilon\ \sin\varepsilon)$ is in the order of magnitude of 1 or less. In practice, $\gamma \approx 1$.

Similar to what we saw in Section 3.1.2, the term I/h (the I-term) in Equation 3.69 and 3.70 is generally small compared to the convergence term γ/b. The average water level slope can be scaled at less than h/L, with L being the length of the tidal influence (the length of the estuary). Hence the ratio of the I term to the convergence term is less than b/L. In estuaries with a strong topography where $b \ll L$, the I term can be disregarded, as for example in the Schelde, an estuary with a strong topography ($b = 28$ km and $L = 200$ km), $b/L = 28/200 = 0.14$. The Incomati in Mozambique is an example of an estuary with a strong topography in the lower segment but with moderately converging banks upstream, where the ratio in the lower segment is $b/L = 6/100 = 0.06$, whereas in the upper segment it is $b/L = 42/100 = 0.42$, thereby implying that the I term may become important in that part. Since the residual slope I is determined by the bottom slope, the effect of the I term may become important in the upstream part of an estuary.

The equations can be simplified by the introduction of three parameters: the dimensionless tidal range y (Equation 3.19), the dimensionless Tidal Froude number α (Equation 3.20), and the tidal damping scale β (Equation 3.21). As a result of the river discharge, the damping scale β needs to be adjusted:

Zone I: $\upsilon \sin\varepsilon > Q_{\mathrm{f}}/A$

$$\frac{1}{\beta_{\mathrm{I}}} = \frac{\gamma}{b} - \frac{f''\upsilon\sin(\varepsilon)}{\bar{h}^{1.33}c_0}\left(1 + \frac{2.66 y \eta_0}{\bar{h}}\frac{Q_{\mathrm{f}}}{A\upsilon\sin(\varepsilon)} + \frac{Q_{\mathrm{f}}^2}{A^2\upsilon^2\sin^2(\varepsilon)}\right) \tag{3.73}$$

Zone II: $\upsilon \sin\varepsilon < Q_f/A$

$$\frac{1}{\beta_{II}} = \frac{\gamma}{b} - \frac{f''\upsilon\sin(\varepsilon)}{\bar{h}^{1.33}c_0}\left(\frac{1.33y\eta_0}{\bar{h}} + 2\frac{Q_f}{A\upsilon\sin(\varepsilon)} + \frac{1.33y\eta_0}{\bar{h}}\frac{Q_f^2}{A^2\upsilon^2\sin^2(\varepsilon)}\right) \tag{3.74}$$

It can be seen from Equations 3.73 and 3.74 that $\beta_I = \beta_{II}$ if $\upsilon\sin\varepsilon = Q_f/A$. Substitution of these new parameters in Equations 3.69 and 3.70 yields for zone I:

$$\frac{dy}{dx}\left(\frac{1}{\alpha} - \frac{\eta_0}{\bar{h}}\frac{Q_f}{A\upsilon\sin(\varepsilon)} + \frac{\gamma}{y}\right) = \frac{1}{\beta_I} - \frac{I}{\bar{h}} \tag{3.75}$$

and for zone II:

$$\frac{dy}{dx}\left(\frac{1}{\alpha} - \frac{\eta_0}{\bar{h}}\frac{Q_f}{A\upsilon\sin(\varepsilon)} + \frac{\gamma}{y}\right) = \frac{1}{\beta_{II}} - \frac{I}{\bar{h}} \tag{3.76}$$

The Q_f-term in the left hand member and $1/\alpha$ can be merged into a single adjusted Tidal Froude number term α', yielding:

$$\frac{dy}{dx}\left(\frac{1}{\alpha'} + \frac{\gamma}{y}\right) = \frac{1}{\beta_I} - \frac{I}{\bar{h}} \tag{3.77}$$

with:

$$\frac{1}{\alpha'} = \frac{1}{\alpha}\left(1 - \frac{c}{g\bar{h}}\frac{Q_f}{A}\right) = \frac{1}{\alpha}\left(1 - \boldsymbol{F}\frac{Q_f}{\upsilon A}\right) \approx \frac{1}{\alpha} \tag{3.78}$$

The main difference between Equation 3.77 and the earlier derived Equation 3.22 (besides the introduction of γ and I) is the introduction of $Q_f/(A\,\upsilon\sin\varepsilon)$ in β and the use of the Manning equation in β.

It can be seen from Equation 3.78 that, since in alluvial estuaries the Froude number $\boldsymbol{F} \ll 1$, the impact of the river discharge on the adjusted Tidal Froude number (α') is small, as long as $Q_f/(\upsilon A)$ is in the order of magnitude of 1 or less. If the I-term on the right hand side is disregarded and if $Q_f = 0$ (implying that: $\gamma = 1$, $1/\alpha' = 1/\alpha$ and $1/\beta_I = 1/\beta_{II} = 1/\beta$), then Equation 3.77 is the same equation as Equation 3.22 which has a linear and a logarithmic part.

In contrast to Equation 3.22, Equations 3.75 and 3.76 have no analytical solutions because $1/\beta_I$ and $1/\beta_{II}$ are functions of x. However Equation 3.22 can be used to assess the relative importance of changes in α and β on the tidal damping. The tidal Froude number α only affects the linear part of the equation, whereas the damping scale β affects both components. Savenije (2001a) showed that, during tidal amplification, the linear part of the equation is dominant over the exponential part and that during tidal damping the exponential part gradually gains importance, particularly when the tidal range becomes very small. Hence in the upstream part, the influence of the river discharge is only felt through β. In the following section, the influence of the river discharge on tidal damping through β is analyzed for the Schelde estuary.

3.3.4 Application to the Schelde-estuary

With Equations 3.75 and 3.76, an analytical model can be made that can be compared to observations made in the Schelde on 14 and 15 June 1995. The characteristics of the model are based on the Schelde geometry. The length of the estuary $L = 200$ km, the width at the estuary mouth $B_0 = 26$ km and the convergence length $b = 28$ km. Up to 110 km from the mouth the depth of flow $h = 10.5$ m. There is no bottom slope for upto 110 km from the mouth, after which the depth reduces gradually to 2.6 m. The estuary is subject to a harmonic tide at the mouth of the estuary ($x = 0$) with a tidal range of $H_0 = 4.4$ m (corresponding with spring tide), a tidal velocity amplitude of $\upsilon = 1.2$ m/s and a tidal period of $T = 44,400$ s. The tidal velocity is amplified and damped in agreement with the damping and amplification of the tidal range, as observed by Graas (2001).

At the upstream end of the model ($x = 200$ km) there is a weir barrage where a river discharge of 38 m^3/s passed downstream on 14 and 15 June 1995. This weir corresponds to the weir in Gent. The total river discharge amounted to 112 m^3/s since there was an additional 74 m^3/s coming from the Rupel tributary at $x = 128$ km.

A phase lag of 40 min has been used throughout the estuary in accordance with observations made by Graas (2001). For the celerity of propagation, the average between HW and LW has been taken, corresponding with the mean tidal situation.

Since it is not possible to solve the differential Equations 3.75 and 3.76 analytically, a numerical equation has been used: $y_{n+1} = y_n + \mathrm{d}y/\mathrm{d}x^* \Delta x$, with a length step $\Delta x = 2.5$ km. The equation can be solved simply in a spreadsheet.

In Figure 3.5, the results of the spreadsheet model are compared to the observed water levels in the Schelde during dead tide on 21 June 1995 ($Q_f = 41$ m^3/s). The model fits the observations using a constant Manning coefficient $K = 38$ $m^{0.33}$/s (n = 0.026) along the estuary axis. This implies that in the lower part of the estuary a Chézy roughness of 57 $m^{0.5}$/s applies. One may conclude that by the introduction of the river discharge in Savenije's (2001a) model, an almost perfect fit with observations has been obtained. In Figure 3.5, the new model is compared to the original model ($Q_f = 0$) and to a situation of high river discharge ($Q_f = 100$ m^3/s). The effect of the river discharge on the tidal damping is considerable in the upstream part. It appears that the tidal range close to Gent is substantially reduced by the river discharge. This is primarily due to the LW levels being higher. The HW levels are less affected by the river discharge.

It can be concluded that β is strongly affected by river discharge if Q_f/A approaches $\upsilon\sin\varepsilon$. This also may have a significant impact on the damping term D in the Celerity equation (Equation 3.51), which is directly proportional to $1/\beta$.

The influence of river discharge on wave celerity is therefore primarily felt through the effect of river discharge on tidal damping. Finally, the deviation seen in Figure 3.10 between the observed and computed wave celerity for HW and LW in the upper reach of the estuary may have three causes:

1. the tidal amplitude-to-depth ratio at LW approaches unity in the area near the 150 km mark; however, the ratio of the tidal amplitude to the average depth remains below 0.5.
2. the ratio of river discharge to tidal flow approaches unity near the 180 km mark.
3. the shift in ε causes the celerity near HW to be close to c_0 and the celerity near LW to be slower.

It is not completely clear which of these effects is most important. It is a fact however, that the analytical solutions become less applicable as we move further upstream and the estuary gradually gains a riverine character.

3.3.5 Conclusion

In the upper reach of the Schelde estuary, there appears to be an important influence of the river discharge on the tidal range. The river discharge is largely responsible for the considerable tidal damping that occurs upstream. The effect of the river discharge on tidal damping is primarily through the friction term. An important point along the estuary is the point where the two moments of slack occur at the same time, upstream of which the tidal flow no longer changes direction and where the river discharge becomes dominant over the tidal flows. At this point, which varies with river discharge, the friction term is dominated by the river discharge. The reduction of the tidal range is primarily caused by higher water levels at LW, as a consequence of the river discharge forcing itself through a narrow cross section.

In this section, an equation has been presented that accounts for the effect of the river discharge on tidal damping and tidal wave propagation. By the introduction of the river discharge into the derivations, a considerable improvement of the existing analytical equation for tidal damping could be obtained. The comparison of the equation with observations is quite good. The equation is not complicated and can be easily applied e.g. in a spreadsheet. Although the equation enhances our insight into the effect of river discharge on tidal damping and propagation, we cannot use the equation far beyond the point where the river discharge and the tidal flow are of equal magnitude.

3.4 THE INFLUENCE OF CLIMATE CHANGE AND HUMAN INTERFERENCE ON ESTUARIES

The equations presented in the previous sections provide us with a very useful tool to assess the possible impacts of human interference in the estuarine system as well as the effects of climatic change. Man and climate can impact on the estuary in a number of ways:

- dredging and deepening of access and shipping channels
- bank stabilization, canalization and constrictions
- closure of tidal branches and inlets
- construction of harbors
- sea level rise
- changed rainfall and evaporation patterns
- modified river discharge regime

Dredging can have a large impact on tidal hydraulics, particularly if it leads to general deepening. If dredging is done in a way that the spill is dumped elsewhere in the cross section, then dredging does not lead to an increase of the cross-sectional average depth. If the spill is moved out of the estuary, however, the average depth increases, which has various implications. There is a difference in the short-term and long-term reaction. The long-term reaction is a morphological reaction which may change the shape of the estuary and particularly, the convergence length. Such a morphological reaction is slow and would most probably be counteracted by engineering works of bank stabilization, the estuary being a focus of engineering attention already. The short-term reaction can be seen from both the Scaling equation (Equation 2.92), the Damping equation and the Celerity equation (Equation 3.51). As the depth increases, the Scaling equation suggests that the tidal velocity would decrease and the celerity increase. The tidal range is fixed by the downstream boundary, but could increase if a shallow sill near the estuary mouth is removed. These changes all point towards decreasing friction in the Damping equation, leading to either reduced tidal damping or increased tidal amplification. As a result, the wave celerity increases. What happens to the Wave-type Number ($\sin\varepsilon$) is not very clear. We can see form the Phase Lag equation (Equation 2.88) that the increase in both c and δ counteract each other, leading to a minor change in ε, if at all. We also see in the Geometry-tide relation that the deepening and the reduced damping (or increased amplification) counteract each other, which leads to a more or less unchanged tidal range to tidal excursion ratio, suggesting that the tidal velocity amplitude remains more or less the same. Both the tidal range and the tidal excursions, however, will amplify in upstream direction.

Bank stabilisation affects the storage width ratio. Fixed banks and closure of tidal creeks and inlets, often in combination with dredging, leads to less storage on banks on tidal flats and in creeks resulting in a value of r_S close to unity. The celerity equation shows that a reduced value of r_S directly leads to a higher velocity of propagation, being inversely proportional to the square root of r_S. As a second order effect, the celerity increases even further because the convergence term in Equation 3.51 increases compared to the friction term. A higher wave celerity in the Damping equation leads to less damping and more amplification.

The Phase Lag equation again provides negative feedback. An increased wave celerity and increased tidal damping counteract each other in the Phase Lag equation, yielding a more or less constant phase lag. In the Scaling equation, the celerity increases with the square root of r_S, so the tidal velocity amplitude (and hence the tidal excursion) is expected to increase at a similar rate. So in conclusion, bank stabilization leads to higher tidal velocity, higher wave celerity and a longer tidal excursion.

Constrictions primarily have local influence. A constriction imposed by bank stabilization, as is the case for instance in the Schelde near the city of Vlissingen (Flushing), leads to channel deepening, while maintaining the cross-sectional area. This is mainly a local effect that does not have a noticeable impact on the overall hydraulic behavior. Canalization has a larger impact. Besides affecting the storage width ratio, it may also change the convergence, as happened with the Rotterdam waterway. In canals with a long convergence length, the Wave-type number will approach unity (progressive wave), leading to tidal damping and a strong tidal velocity gradient. Construction of harbors on tidal channels will lead to more storage, reduced wave propagation and more tidal damping. In principle, the loss of storage width caused by dredging and vertical walls can be compensated by the gain in storage through harbor construction.

The possible impact of sea level rise is a topical issue. Over the next century, the rate of sea level rise could be in the order of 0.5 m, but it could also be more. Let us assume that sea level rise will be accompanied by the raising of estuary banks, and hence loss of storage width. This implies that sea level rise will be much the same as a combination of deepening and storage width reduction. On top of that, sea level rise may increase the tidal range at the downstream boundary, particularly in estuaries that have a shallow sill near the mouth (such as the Incomati). The combination of these effects, which strengthen each other, leads to a higher wave celerity, more tidal amplification, and a larger tidal excursion. Therefore, in addition to the sea level rise, people living along an estuary will have to reckon with a larger tidal range (so higher HW) and a shorter travel time of the tidal wave. How much this effect will be depends strongly on the characteristics of the estuary. Because of the non-linearity of the Damping equation, the reaction of the estuary system very much depends on the values of the hydraulic parameters. With a simple spreadsheet model however, the combination of the equations presented in Table 3.3 can provide good indications of what should be expected if certain changes are made to the geometry and the hydrological boundary conditions of an estuary.

Finally, there is the effect of hydrology and climate. We have seen that the river discharge affects the hydraulics of the estuary, particularly the tidal damping and the wave celerity in the riverine part of the estuary. These effects are not as dramatic as the ones discussed above. However, the impact of climate and hydrology on salinity, water quality, and ecosystem behavior can be substantial, possibly leading to drastic changes in overall system behavior. These effects will be discussed further in Chapter 4.

4

Mixing in alluvial estuaries

In well-mixed estuaries, salinity penetrates through the process of mixing while the river discharge flushes it back towards the sea. This struggle for dominance can have two winners: the mixing, and in that case we observe an increasing salinity over time; or the river flow, and then we see the estuary water become fresher. When the two mechanisms tie, we have a steady-state situation where the salinity remains constant over time. What remains is a longitudinal gradient of the salinity, gradually diminishing in upstream direction from sea salinity at the mouth to fresh water at the toe of the salt intrusion curve.

There are several mixing mechanisms that vary in importance depending on the shape of the estuary, the location, the level of stratification, the density, and the strength of the tide. One can distinguish different types of mixing, such as mixing by turbulence, mixing by tidal shear, mixing by residual currents, mixing by trapping, and density-driven mixing. These mechanisms will be described in the following sections. All these mixing mechanisms together drive longitudinal dispersion of salinity, which can be decomposed into many smaller constituting fluxes. There are different methods for flux decomposition, but we shall see that this approach does not really lead to practical results. In order to obtain a predictive model, we require a predictive equation for the effective longitudinal dispersion. This one-dimensional (1D) predictive equation will be derived and illustrated by empirical data. A general equation that integrates all mixing processes will be presented.

4.1 TYPES OF MIXING, THEIR RELATIVE IMPORTANCE, AND INTERACTION

Mixing is the mechanism through which salt travels upstream. During every tidal cycle, on the flood tide, an amount of salt water enters the estuary, but if that amount of water does not mix, then the same water again leaves the estuary on the ebb tide without the salinity penetrating further. We shall see that if we want to analyze mixing in detail by looking at all the different mixing mechanisms at their particular spatial and temporal scales, the picture becomes very fuzzy. Several authors have tried to split up the mixing process into smaller components resulting from spatial and temporal averaging, but without enhancing the insight into how mixing works. Jay et al. (1997) concluded that the track record of determining these fluxes is discouraging, among others because of the low accuracy that can be reached in subtracting fluxes. Others have looked at the driving mechanisms of mixing and what this teaches us about the main hydraulic parameters that influence mixing. Although this approach enhances our insight, to date, it has failed to come up with a predictive method to forecast the effective dispersion. The most important reason being that we do not know how these individual mechanisms interact and how they provide feedback on each other.

In analogy with Sivapalan et al. (2003), we may call this approach, where we try to build-up the dispersion from analyzing the detailed mixing processes, a 'bottom-up' approach. Like in hydrology, this 'bottom-up' approach, although physically appealing, does not generate workable models that predict system behavior. The reasons lie partly in a phenomenon called 'equifinality' and partly in the fact that these individual components do not function independently but interact according to certain laws of 'self-organization.' In tidal dispersion similar processes are at work. Therefore, it is worthwhile to look at the concept of equifinality and self-organization in some detail.

The concept of equifinality is notorious in hydrology. It was introduced by Beven (1993) to describe the fact that distributed rainfall–runoff models may perform well, but often for the wrong reasons. Distributed physically based rainfall–runoff models use large sets of spatially distributed parameters. It appears that the same hydrological behavior can be simulated adequately by a sheer infinite combination of parameters, which often are not even close to their expected value. As a result, these complex hydrological models cannot do much more than mimic hydrological behavior, but their predictive value, to forecast what would happen if we changed something in the land use of the catchment, is low. At first sight this is a disappointing result. It implies that a purely physically based approach of looking at sub-processes at detailed scale, and subsequently scaling these up to the larger scale, does not yield satisfactory results. On the other hand, equifinality is an indication of the existence of a physical law that apparently translates a plethora of detailed processes (in a strongly heterogeneous environment) into consistent system functioning. As a result, Savenije (2001b) called 'equifinality, a blessing in disguise.' Equifinality is the reason why relatively simple hydrological laws exist, which are able to describe hydrological processes under highly variable conditions, in different physical environments and in far from homogeneous situations. Coming to grips with the underlying physical law that govern equifinality is one of the biggest challenges of hydrology. And the same may be true for mixing in estuaries.

The main question is: what causes equifinality? One important cause is that hydrology is a complex system of interacting processes that provide feedback on each other while attenuating extremes. The physical process underlying this is entropy. To say it popularly, entropy does not like extremes. The second law of thermodynamics implies that average behavior becomes overwhelmingly likely in a very large system, implying that exceptions will no longer be observable in the output signal of large systems. In the end all energy is transferred into heat. Within a large system, this energy is dissipated as evenly as possible.

A watershed is a large complex system, and so is an alluvial estuary. Just like an alluvial estuary, a watershed shapes the medium through which the water flows. Erosion, deposition, and biological activity are the main shaping forces, but the underlying physical law is the maximization of entropy driving self-organization in a way that energy is dissipated as homogeneously as possible. Friction is the most important force that translates energy into heat. Different processes interact to spread the energy smoothly over the water trajectory. If one process is overloaded it triggers another. The feedback between these processes results in an overall system performance obeying a physical law at a higher level of aggregation, and there are many ways that this behavior can be reached.

Mixing works the same way. Mixing is spreading of energy. If one mixing mechanism is under performing, another takes over; not because they communicate, but because the physical processes that shape the geometry, and that drive the mixing processes, are connected by the second law of thermodynamics. Hence there are feedback mechanisms that lead to efficient and gradual dissipation of mixing energy. The law to describe this overall mixing behavior still has to be derived, but in this chapter a formula will be presented (Van den Burgh's equation) that comes very close to it, judging from its excellent performance and closeness to the theoretical knowledge available to date.

Following the analogy of Sivapalan, Van den Burgh's method is a 'top-down' approach, which is mainly empirical and based on what the data 'tell us.' We observe certain system behavior, we derive an equation that describes it and we relate the key parameters of the equation to the physical parameters that we know are the main drivers of the process. This is the way that many physical laws have been discovered: the Gas law, Darcy's law, Manning's law, Newton's law of gravity, etc. There is nothing wrong with it. The only problem is that we feel uncomfortable if we cannot make the connection between what we observe at a small scale and what we observe at the aggregated system-scale. We feel dissatisfied, as with the magician who just fooled us into believing that the girl has been cut into two and we have no clue how the trick works. At the same time it triggers our curiosity. Many physical scientists are still trying to work out how gravity works. Yet nobody contests Newton's law describing it.

It is clear that for a 'top-down' empirically derived physical 'law' to be credible, it has to be (1) based on solid empirical evidence in a wide range of situations, (2) consistent with other certified physical laws, (3) based on the dominant physical drivers that we know, and (4) connected to the physical processes we observe at smaller scales. The last condition implies that we bring the 'top-down' and 'bottom-up' approaches together, similarly to what is advocated by Sivapalan. So let us look at the mixing mechanisms and at what drives them. Although detailed study of the mixing processes may not be the right way to understand system performance, we have to understand them to find the middle ground between the top-down and the bottom-up approaches.

There is virtually no limit to the number of mixing processes that can be identified. Fischer et al. (1979) separated the small-scale turbulent diffusion (periodicity less than a few minutes) from larger scale advective processes, although the separation between the two is arbitrary. Turbulence essentially is the mechanism that transfers the friction from the estuary/river bottom into the body of the flowing water. Gravity works on all water particles, but the friction only along the interface between land and water. The sheer stress exercised on the interface is transferred into the fluid by turbulent eddies that dissipate energy within the fluid. This causes mixing by turbulent eddies at spatial scales of a few meters and timescales of less than a few minutes, but also interactions at larger scales. The flow can be considered to exist of different streamlines that flow in different directions and at different velocities. There is a shear stress exercised between these stream lines which we call tidal shear. Where streamlines interact, cross-over, or meet to exchange fluid, we talk of advective dispersion. An essential difference between a river and an estuary is that the magnitude, the direction, and even the existence of these streamlines is continuously changing over time and space, as a result of tidal forcing. The direction of the flow lines in an estuary is seldom parallel to the estuary axis, but is shearing between flood and ebb channels. This makes the mixing highly dynamic.

Besides tidal-forced mixing there is also mixing by wind and by the river. Hence, we can distinguish three main driving forces for mixing:

- the wind that drives both vertical and horizontal circulation. The vertical circulation is driven by wind shear inducing a surface current of relatively fresh water and a water-level slope in the direction of the wind, while the surface slope triggers a relatively saline return flow close to the bottom (see Figure 4.1). Mixing occurs along the interface between these two currents and through upwelling of relatively saline water from the bottom. The wind also can cause horizontal circulation depending on the shape of the estuary. Particularly irregular estuaries, such as Rias, can experience net circulation over shallow bays due to wind (see Fischer et al., 1979). Although in lakes and coastal lagoons wind-driven mixing can be dominant, in alluvial estuaries this mixing mechanism is considered less important than the following two.
- the river provides a deficit of potential energy with buoyant fresh water driving vertical gravitational circulation. Gravitational circulation is an important mechanism in the part of the estuary where the longitudinal salinity gradient is the largest. In estuaries with a strong funnel shape (and hence a dome-shaped salt intrusion curve), this region is located in the central part of the salt intrusion length. In the downstream part of these estuaries, where the salinity gradient is small, tide-driven mixing is dominant. In narrow estuaries, with a recession-shaped salt intrusion curve and a rather constant salinity gradient, gravitational circulation is the main mixing mechanism throughout.
- the tide provides kinetic energy to the estuary that can overcome the potential energy deficit of the river water. The tide rocks the estuary water back and

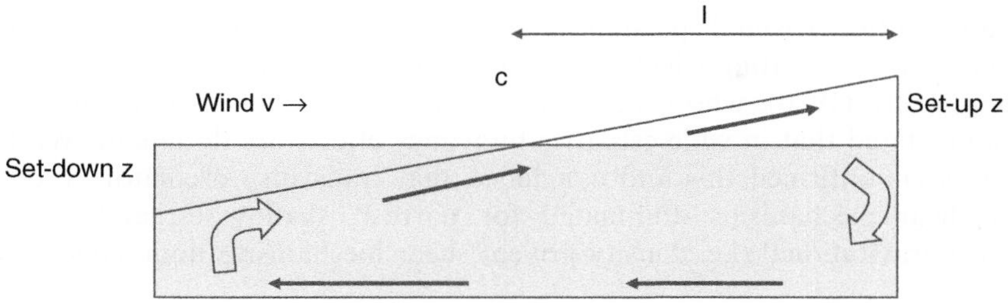

Figure 4.1 Wind-driven vertical circulation.

forth and dissipates tidal energy through mixing. The tide generates different types of mixing: (1) turbulent mixing at small spatial and temporal scales; (2) tidal shear between streamlines with different velocities; (3) spring–neap interaction; (4) trapping of water on tidal flats and in dead ends; (5) residual currents in the cross section; (6) residual currents over tidal flats and shallows; (7) exchange between ebb and flood channels that meet and mix at cross-over points. The latter mechanism is dominant in the downstream part of estuaries with a dome-shaped salt intrusion curve.

We saw earlier, in Chapter 1, that the balance between the potential energy deficit and the tidal kinetic energy is reflected in the Estuarine Richardson number N_R (introduced by Fischer, 1972), which is represented here:

$$N_R = \frac{\Delta\rho}{\rho}\frac{gh}{\upsilon^2}\frac{Q_f T}{P_t} \tag{4.1}$$

Hence the Estuarine Richardson number is a measure for the relative importance of gravitational circulation compared to tidal mixing. The ultimate form of gravitational circulation is the saline wedge, which corresponds with a high Estuarine Richardson number.

In the following sections, we shall discuss the mixing by the tide and by the river in more detail. We shall pay no further attention to wind-driven mixing and concentrate on the main mechanisms. Gravitational circulation and mixing by tidal shear have been well documented in the literature. These mechanisms will be briefly summarized. Still poorly known are the mixing mechanisms by tidal pumping and residual circulation. Here we shall try to explore new terrain.

4.2 GRAVITATIONAL CIRCULATION

Hansen and Rattray (1965) started a discussion on which of the two mechanisms (density-driven or tide-driven) is dominant in a certain estuary. Their classification method makes use of the parameter ν which reflects the relative importance of tide-driven dispersion *versus* the total dispersion. Several researchers spent time on investigating which of the two is the dominant mechanism in certain estuaries. Smith (1980), who called density-driven dispersion as buoyancy-driven dispersion, stated that in wide estuaries buoyancy effects are dominant. West and Broyd (1981) confirmed this and concluded that transverse oscillatory (i.e. tide-driven) shear mechanisms dominated for narrow, shallow estuaries; whereas transverse gravitational (i.e. density-driven) shear mechanisms dominated in wide estuaries.

Here we come to a different conclusion. Gravitational circulation is dependent on the longitudinal salinity gradient and driven by the moment M per unit of volume that results from the two opposed hydrostatic forces (of the fresher upstream section and the more saline downstream section) that are equal in magnitude but do not work along the same line of action (see Figure 2.2). Equation 2.32 that describes the moment exercised on the fluid is represented here:

$$M = \frac{1}{12}\frac{\partial \rho}{\partial x} g h^2 \tag{4.2}$$

We see that the longitudinal salinity gradient is not large in wide estuaries because wide estuaries (with a short convergence length) have a dome-shaped intrusion curve, which has a very small salinity gradient in the wider part of the estuary. Only as the estuary narrows does the salinity gradient become stronger. The observations by earlier researchers are often biased by two shortcomings. The first is that much work has been done on the basis of laboratory flumes or mathematical models that assume constant width (e.g. Hansen and Rattray, 1965). These results are of limited use in real estuaries with converging banks. The second is that if observations were made in real estuaries, they were merely done in selected cross-sections of the estuary (often only one), which may not be representative at all for the overall longitudinal dispersion.

Instead we see that gravitational circulation is dominant in near prismatic estuaries (with a long convergence length), such as the Limpopo, the Chao Phya, or the Rotterdam Waterway, experiencing a steep salinity gradient, but which are generally narrow compared to estuaries with a short convergence length.

The apparent paradox between what e.g. Smith (1980) observed (that gravitational circulation is larger in wide estuaries) and what is stated here can be explained if we distinguish between width and convergence. An estuary with a short convergence length is wide at the mouth. If we look at the estuary from the upstream end, then the width increases gradually from the river width to the sea. The shorter the convergence length, the wider the estuary becomes. Because the tidal influence in alluvial estuaries always exceeds a quarter of the tidal wave length, which is in the order of 100 km or more, estuaries with a short convergence length are always wide. Similarly estuaries with a long convergence length are generally narrow, having a width not much larger than the river width. We saw that estuaries with a short convergence length (i.e. wide estuaries) are dominated by tidal mixing, and estuaries with a long convergence length (i.e. narrow estuaries) by gravitational circulation. In the wider part of the estuary the salinity gradient is small and hence the gravitational circulation is small. But if there is a salinity gradient then we can indeed conclude that the wider an estuary is, the more powerful the gravitational circulation will be.

Fischer et al. (1979) observed that the density gradient drives lateral mixing rather than vertical mixing, as a result of the varying depth over the cross section. Because estuaries are much wider than deep, lateral gravitational circulation is much more important than vertical circulation. Overall effective longitudinal dispersion by gravitation circulation D_g, according to Fischer, is proportional to the width squared and the depth to the sixth power:

$$D_g \propto \left(\frac{g}{\rho}\frac{\partial \rho}{\partial x}\right)^2 h^6 B^2 \tag{4.3}$$

Hence we see that this gravitational dispersion is both a function of the width and the salinity gradient. Both should be significant for this type of dispersion to be dominant. An estuary that is wide, but has no salinity gradient will not experience much gravitational circulation.

Fischer concludes by saying that although we have started to understand the mechanisms at work in tidal and gravitational mixing, we must be very cautious to apply formulae that have been derived under laboratory conditions or on the basis of spot observations in real estuaries. *'We have given several formulas for estimating the value of the longitudinal dispersion coefficient, but each one has been based on an analysis of one mechanism at the neglect of others.'* It is especially interesting to find out how these mechanisms interact. One of the few articles that deal with the combined effect of gravitational circulation and tidal dynamics is by McCarthy (1993), which will be briefly presented below where we discuss residual circulation.

4.3 MIXING BY THE TIDE

We saw that the tide generates different types of mixing which we shall describe in somewhat more detail below.

Turbulent mixing is the weakest of the mechanisms occurring at small spatial (a few meters) and temporal scales (a few minutes). Fischer et al. (1979) consider it inferior to the other tide-driven mechanisms that can be classified as advective dispersion. The latter results from water flowing in streamlines that move at different velocities, in different directions, and that vary over time. These streamlines interact and exchange fluid. A practical distinction between turbulent and advective mixing is that a three-dimensional (3D) hydraulic model is able to model the salt fluxes resulting from advective dispersion by the combination of the velocity field with the salinity field. The turbulent dispersion is imposed through the eddy diffusivity of turbulent flow that these models use. Hence a good three-dimensional hydraulic model should be able to simulate tidal mixing adequately.

Uncles and Stephens (1996) emphasized the importance of spring–neap interaction. During neap tide, estuaries tend to be more stratified, as the Estuarine Richardson number is larger. The strength of tidally driven mixing may vary significantly between spring and neap tides (Jay and Smith, 1990b). The transition from neap to spring tide can generate significant mixing.

Schijf and Schönfeld (1953) introduced the concept of tidal trapping. Tidal trapping results from the phase difference between the main estuary branch and a dead-end tidal branch, bay or tidal flat. In a dead-end branch, slack occurs at HW, whereas the water in the estuary is still flowing upstream at HW. Between HW and HWS the water level drops and the dead-end branch already starts emptying while the estuary still flows upstream with relatively saline water. Hence a tidal flat discharges relatively fresh water into the flood flow. In estuaries with an irregular topography trapping can be an important mechanism. Because trapping occurs only along the sides of the estuary, its relative importance is less in very wide estuaries. The typical length scale of tidal trapping is the tidal excursion E.

A phenomenon receiving more attention in recent years is residual circulation in the cross section. Unfortunately, also here, most of the research has been done on estuaries with constant cross section and 2D vertical mathematical models (e.g. Li and O'Donnell, 1997), or on observations in a single cross section (e.g. Jay and Smith, 1990a; Turrell et al., 1996; Stacey et al., 2001). McCarthy (1993) is an exception. He presented one of the very few articles on residual circulation generated by the combined effect of tide and gravitational circulation in an estuary with exponentially varying width. He used a 2D vertical model and perturbation analysis to identify the mixing mechanisms that combine into longitudinal dispersion. In the estuary with exponentially varying width, McCarthy indeed obtained a dome-shaped intrusion curve and hence a very slight density gradient near the mouth. He concluded that density-driven mixing is weak at the estuary mouth and tidal-induced landward buoyancy transport is dominant. Further inland, the density-driven mixing takes over to counteract the seaward Lagrangean advection of salt. The density-driven mixing is a function of the salinity gradient, whereas the tide-driven mixing is rather a function of the salinity and the width.

Finally there is the type of residual circulation not considered by McCarthy (1993), which Fischer et al. (1979) call 'tidal pumping.' It is partly the result of an irregular topography (as is prominent in Rias but not in alluvial estuaries) and partly of the existence of separate ebb and flood channels that have cross-over points. The latter is a dominant mechanism in the wider part of funnel-shaped estuaries and is discussed in the next section.

4.4 RESIDUAL CIRCULATION THROUGH FLOOD AND EBB CHANNELS

Strongly funnel-shaped estuaries develop separate flood and ebb channels. The Schelde presented in Figure 4.2 is a good example, but similar patterns can be observed in other funnel-shaped estuaries such as the Pungué, the Columbia,

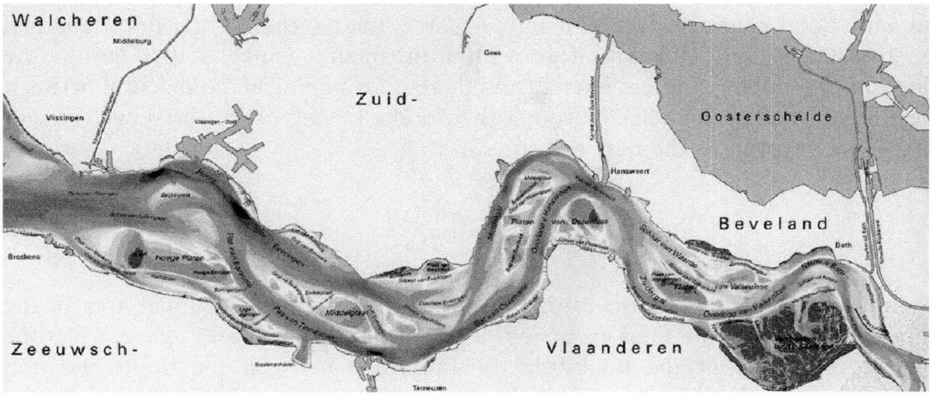

Figure 4.2 Flood and ebb channels in the Schelde.

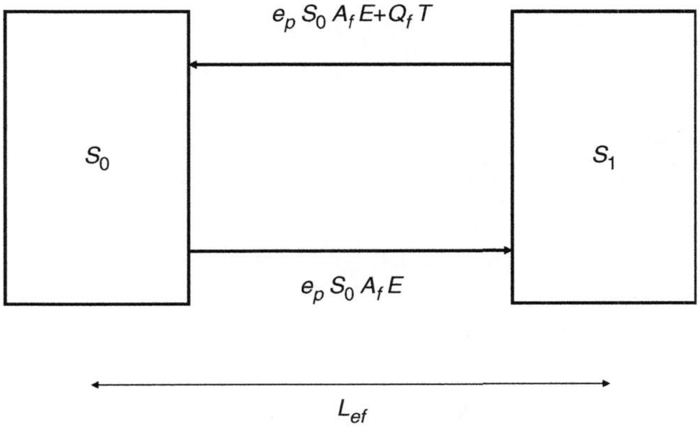

Figure 4.3 Box model for ebb–flood channel dispersion.

and the Thames. In the flood channel, the amplitude of the landward tidal velocity is about 20 percent larger than in the ebb channel. Also the flood channel is about 20 percent shorter than the ebb channel. As a result, on the incoming tide, the relatively saline water flowing through the flood channel arrives earlier at the cross-over point than the relatively fresh water in the ebb channel. On the ebb tide, the amplitude of the tidal velocity in the ebb channel is substantially higher than in the flood channel, about 40 percent. Although the ebb channel is 20 percent longer than the flood channel, the relatively fresh water flowing through the ebb channel reaches the cross-over point earlier than the relatively saline water ebbing through the flood channel.

In the following, a relation is derived for the salt dispersion by this type of residual circulation based on a box model, see Figure 4.3. The box model represents one ebb–flood channel loop of length L_{ef}. We assume that in the flood channel the transfer of salt is Lagrangean with little mixing (since at the mouth the longitudinal salinity gradient is small and there is no significant connection between the ebb and flood channels). Through the flood channel, over a tidal cycle, a salt flux F_f is conveyed to the next cell equal to:

$$F_{\mathrm{f}} = S_0 A_{\mathrm{f}} \frac{E}{T} \frac{\Delta \upsilon}{\upsilon} \tag{4.4}$$

where S_0 is the salinity at the estuary mouth, A_f is the cross-sectional area of the flood channel, E is the tidal excursion, T is the tidal period, and $e_p = \Delta\upsilon/\upsilon$ is the relative difference of the tidal velocity amplitude between the flood and ebb currents in the flood channel, which is the tidal pumping efficiency. In order to close the salt balance, assuming a steady-state situation, this flux should be counteracted by the salt flux F_e in downstream direction through the ebb channel:

$$F_{\mathrm{e}} = -S_1 \left(A_{\mathrm{f}} \frac{E}{T} \frac{\Delta \upsilon}{\upsilon} - Q_{\mathrm{f}} \right) \tag{4.5}$$

where Q_f is the fresh water discharge, which is negative because the positive x-axis points upstream. The sum of these fluxes should be zero, yielding:

$$(S_1 - S_0) A_{\mathrm{f}} \frac{E}{T} \frac{\Delta \upsilon}{\upsilon} = Q_{\mathrm{f}} S_1 \tag{4.6}$$

or:

$$\frac{\partial S}{\partial x} L_{\mathrm{ef}} A_{\mathrm{f}} \frac{E}{T} \frac{\Delta \upsilon}{\upsilon} = Q_{\mathrm{f}} S \tag{4.7}$$

The general steady-state salt dispersion equation is written as:

$$DA \frac{\partial S}{\partial x} = Q_{\mathrm{f}} S \tag{4.8}$$

where D is the longitudinal effective tidal average dispersion coefficient. The derivation of this equation is presented in detail in the next chapter (see, Equation 5.16), but merely represented here. If we compare Equations 4.7 and 4.8 we immediately see the expression for D_{ef}, the effective tidal average dispersion coefficient resulting from residual circulation in the ebb–flood channel system:

$$D_{\mathrm{ef}} = \frac{A_{\mathrm{f}}}{A} \frac{\Delta \upsilon}{\upsilon} \frac{E}{T} L_{\mathrm{ef}} = 0.5 e_{\mathrm{p}} \frac{E}{T} L_{\mathrm{ef}} \tag{4.9}$$

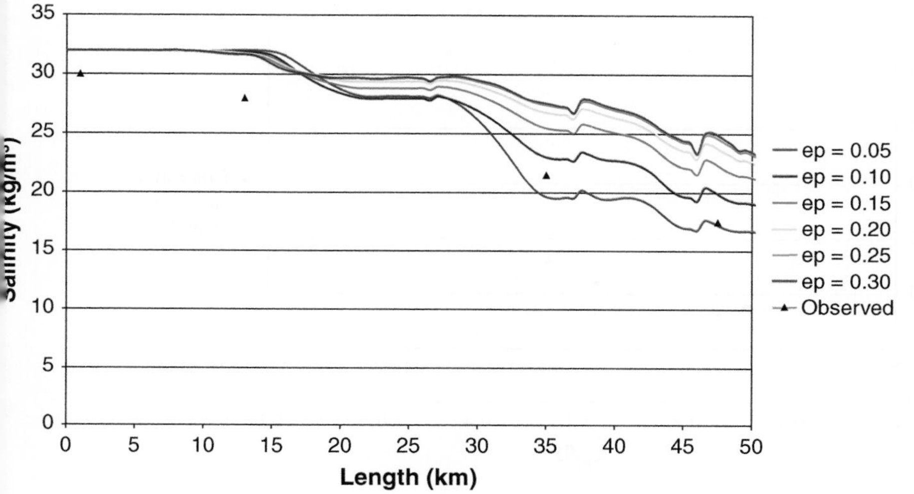

ure 4.4 Lagrangean model for ebb–flood channel interaction considering different nping efficiencies (e_p).

is is a very straightforward and simple result. It implies that salt intrusion due residual circulation between ebb and flood channels is proportional to: the length of ebb–flood interaction loop, 2) the tidal excursion, and 3) a tidal mping efficiency $\Delta v/v$. The ratio of A_f/A may be assumed to be close to 0.5, deviation can be included in the pumping efficiency e_p.

This relationship was tested by Nguyen Anh Duc (unpublished) using a simple grangean model where mixing only takes place at the cross-over points. The del shows that in the most downstream loop pure seawater fills the flood nnel, while mixed, somewhat fresher, water flows down the ebb channel. In the ond loop the same happens, resulting in a decreasing salt intrusion following a iir-case' pattern (see Figure 4.4). The intrusion follows a dome shape. We also that the longer the loop length, the stronger is the dome shape. If the loop gth and the tidal pumping efficiency is increased the longitudinal dispersion reases proportionally, in agreement with Equation 4.8. Subsequently, we studied relationship between the loop length and the estuary geometry, the loop length ng a crucial parameter that appears to become smaller as the estuary becomes rower. Figure 4.2 presents an illustration of the flood–ebb channel pattern in Schelde estuary. The following approach was followed.

Since the loop length scales at the width convergence b, the dimensionless ratio L_{ef} to b was analyzed. It appears that there is a certain width at which separate and flood channels no longer develop. This width B_L depends on the width to pth ratio of a stable channel, so it is logical to assume that there is a fixed ratio B_L/h that forms the threshold for separate ebb and flood channels to develop. the Schelde and the Columbia this ratio lies at about 100. The values of B_L

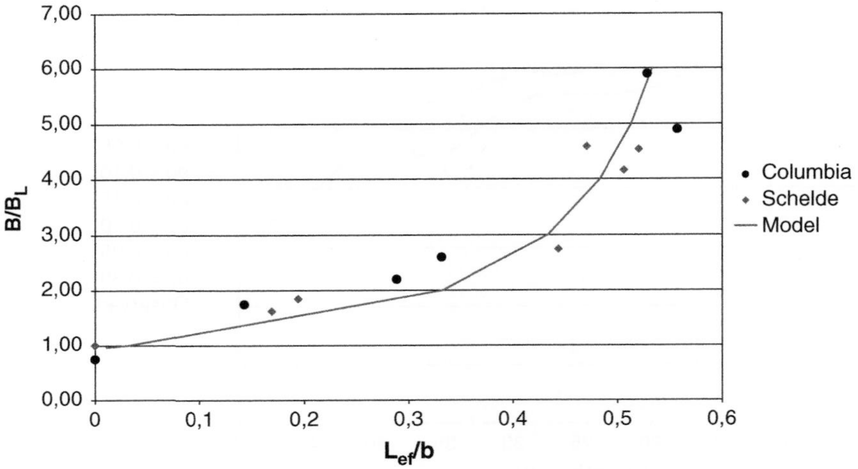

Figure 4.5 Relationship between the loop length to convergence length ratio L_{ef}/b and the dimensionless width B/B_L.

in the Schelde and Columbia are both about 1000 m, the average depth in both estuaries being about 10 m. Now the geometric relationship sought should be a relationship between the loop length and the width at the cross-over point between two loops. One can see the analogy with a standing wave of which the nodes are located at the constriction point where the ebb and flood currents cross-over. The width of the constrictions and the angle of the banks determine the space within which the sinus wave can develop. Figure 4.5 shows a relationship between L_{ef}/b and B/B_L for the Schelde and the Columbia. The points of both estuaries fit the same pattern. The equation describing this relationship is based on the equation derived for the effective longitudinal dispersion presented in Section 4.8.

Because near the mouth of a funnel-shaped estuary the salinity gradient is very small, the dominant mechanism near the mouth is the residual circulation by ebb–flood channel interaction. So near the mouth of funnel-shaped estuaries the dispersion by the ebb–flood channel interaction may be assumed to be equal to the effective longitudinal dispersion. We saw that the dispersion generated by this mechanism is proportional to the length of the ebb–flood channel loop. As a result, the loop length L_{ef} should obey a similar equation as the effective longitudinal dispersion presented in Equation 4.34. The equation, hence has to be of the following type:

$$\frac{L_{ef}}{b} = \alpha_L \left(1 - \beta_L \left(\frac{B_0}{B} - 1\right)\right) \tag{4.10}$$

The condition that $L_{ef}=0$ at $B=B_L$ yields an expression for β_L, resulting into:

$$\frac{L_{ef}}{b}=\alpha_L\left(1-\left(\frac{B_L}{B_0-B_L}\frac{B_0-B}{B}\right)\right)=\alpha_L\left(\frac{1-B_L/B}{1-B_L/B_0}\right) \tag{4.11}$$

This equation indeed fits the data points very well, as we can see in Figure 4.5. It yields a value for α_L of 0.5. In view of the straightforward analysis made the result looks surprisingly accurate. The good fit is a confirmation of the adequacy of the box model and the assumption that ebb–flood channel interaction is the main mechanism in the lower part of funnel-shaped estuaries and it supports the validity of the equation for the effective longitudinal dispersion equation. However, certainly for the time being, we should consider Equation 4.11 as a result obtained by curve fitting rather than as a physical 'law.'

The resulting equation for the dispersion generated by the residual circulation in ebb and flood channels then reads:

$$D_{ef}=\frac{e_p}{4}\frac{Eb}{T}\left(\frac{1-B_L/B_0\exp(x/b)}{1-B_L/B_0}\right) \tag{4.12}$$

This equation is only valid if $B > B_L$. If $B \leq B_L$ then $D_{ef}=0$. If $B > B_L$ we see from Equation 4.12 that at $x=0$ the dispersion due to residual circulation equals $e_p\,Eb/(4T)$. This is a surprisingly simple result.

In summary, we have seen that in funnel-shaped estuaries ebb and flood channels occur if the estuary is wide enough for these channels to develop. In the Schelde and Columbia, this happens downstream from the point where the width to depth ratio is about 100. Downstream from this point, the residual circulation between ebb and flood channels becomes a dominant mechanism (if the salinity gradient is small). The mixing by this mechanism strongly depends on the estuary width, which forces the loop length. The effective longitudinal dispersion generated by the flood–ebb channel interaction is directly proportional to the loop length, the tidal excursion, and the efficiency of the tidal pumping.

Being such a dominant mechanism, it is surprising that so little research has been done on this type of circulation. There may be a number of reasons for this apparent lack of knowledge:

1. To study this mechanism in the field implies a major operation. A survey would involve a dense network of monitoring points and have to stretch over a considerable period of time to monitor the spring–neap interaction.
2. Three-dimensional hydraulic models can reproduce the mixing by flood–ebb channel interaction. There does not appear to be a need for understanding how the mechanism works, if our models can mimic it.
3. The theoretical research on mixing is still too much focused on 2D mathematical modeling and analysis in a single cross section. Or as Jay et al. (1997)

put it: *'estuarine circulation theory has focused on two-dimensional analyses that treat either vertical or lateral variations but not both.'*

4.5 THE DECOMPOSITION METHOD AND WHY IT IS NOT VERY USEFUL

Another way of differentiating between mechanisms is by decomposing the longitudinal salt flux through a cross section into different components. Following the maximum generality scaling approach, Smith (1980) arrived at four dominant mechanisms for well-mixed estuaries the oscillatory vertical shear (earlier described by Bowden, 1967, 1981), the oscillatory transverse shear (earlier described by Okubo, 1967), the interaction between tidal and buoyancy effects, and the buoyancy-driven steady horizontal circulation. Another approach, originally considered by Hansen (1965), was followed by many researchers who arrived at a considerable number of decomposed mixing mechanisms such as vertical and transverse shear dispersion (West and Mangat, 1986); transverse net circulation (tide-driven, density-driven, and boundary-induced, analyzed by Fischer, 1972); vertical net circulation (earlier analyzed by Hansen and Rattray, 1965); transverse (and vertical) oscillatory shear (analyzed by Holley, Harleman and Fischer, 1970); and transverse (and vertical) gravitational net circulation (West and Broyd, 1981). In Hansen's approach, the salinity, the velocity, and cross-sectional area are considered to be the sum of a tidal mean value (subscript 0), a tidally varying (subscript 1), and a turbulent value (subscript 2):

$$s(x,y,z,t) = s_0(x) + s_1(x,t) + s_2(x,y,z,t) \tag{4.13}$$

$$U(x,y,z,t) = U_0(x) + U_1(x,t) + U_2(x,y,z,t) \tag{4.14}$$

$$A(x,y,z,t) = A_0(x) + A_1(x,t) \tag{4.15}$$

The salt flux F can be defined as:

$$F = \int Us\,\mathrm{d}A \tag{4.16}$$

The tidal average salt flux can then be decomposed into six terms (Fischer, 1972):

$$\begin{aligned}\langle F\rangle &= A_0U_0s_0 + \langle A_1U_1\rangle s_0 + A_0\langle U_1s_1\rangle \\ &\quad + U_0\langle A_1s_1\rangle + \langle A_1(U_1s_1)'\rangle + \langle AU_2s_2\rangle\end{aligned} \tag{4.17}$$

in which $(U_1s_1)'$ is the deviation of U_1s_1 from its tidal mean and where the angle brackets denote a tidal average value. The first two terms on the right-hand side constitute the advective salt flux caused by the fresh discharge Q_f (which has a negative value since it points downstream). The advective Lagrangean salt transport resulting of the river discharge $A_0 U_f s_0$ is not equal to the first term only, which is the Eulerian average salt transport, directed downstream. Dyer (1973) indicated that the second term, the so-called Stokes drift, is part of the Lagrangean transport and follows from the river discharge not being equal to the Eulerian integral, as the cross-sectional area varies over time with amplitude $\hat{A}$. This can be seen as follows:

$$Q_f = \frac{1}{T}\int AU\mathrm{d}t = \overline{AU} - \frac{1}{2\pi}\int \hat{A}\upsilon \sin(\omega t - \varepsilon)\cos(\omega t)\mathrm{d}t \qquad (4.18)$$

where the harmonics are chosen in correspondence with the definition for h and U in Chapter 3 (Equations 3.30–3.31). The first term on the right-hand side is the Eulerian mean discharge which is directed downstream, whereas the second term represents the Stokes drift, which is directed upstream (the mean Eulerian velocity is negative). It implies that in absolute terms the Eulerian discharge should be larger than the Stokes drift by an amount equal to Q_f. We can see that the Stokes drift is zero for a standing wave ($\varepsilon = 0$); since the integral of $\sin(2\alpha)$ over 2π is zero. In that case the Eulerian discharge equals Q_f. For a progressive wave, however, with $\varepsilon = \pi/2$, the harmonics are exactly in phase. The argument of the integral then is $-\cos^2\alpha$, of which the integral is $-\pi$. Hence for a progressive wave, the Stokes drift equals $\hat{A}\upsilon = B\eta\upsilon$ (in upstream direction). If we scale the Stokes drift against the Lagrangean salt transport, then the proportion is $\eta\upsilon/hU_f$ (with U_f as the velocity of the fresh water discharge). Making use of the Scaling equation (Equation 2.92) with $\varepsilon = \pi/2$, this implies that this ratio scales as the ratio of the Froude number to the Canter Cremers number: $\boldsymbol{F}/N$. The Froude number is in the order of 0.1 or smaller and does not vary along the estuary axis, but in well-mixed estuaries, the Canter Cremers number can be substantially less than that. It reduces in downstream direction proportional to the cross-sectional area. It is unity at the point where there is only one moment of slack (point P in Figure 1.6 where $U_f = \upsilon$). Near the mouth of a funnel-shaped estuary the ratio becomes very small, and the Stokes drift very large. Because the Lagrangean salt flux remains the same, the Eulerian discharge should increase at the same rate (their sum being equal to the Lagrangean salt flux). So in an alluvial (positive) estuary, where $\varepsilon > 0$, the Stokes drift is always larger than (and opposed to) the Lagrangean salt flux, the balance between them being the Eulerian (non-tidal) discharge.

Van de Kreeke and Zimmerman (1990), following the suggestion of Fischer (1972), split up s_2 and U_2 even further into vertical and transverse components, but neglected tidal variations in A, leading to 6 components containing advection, geometry-induced dispersion, residual lateral circulation, vertical density circulation, lateral oscillatory shear, and vertical shear. Park and James (1990), who (after Dyer, 1974) in addition considered the tidal variation, decomposed the salt flux into 66 components, grouped into an equation of 11 terms. These components had to be grouped to be able to attribute some physical meaning to them. Jay et al. (1997) observed: *'Little attempt has been made to connect estuarine circulation to the salt fluxes that must maintain it. The result has been a welter of confusing transport expansions filled with terms of uncertain meaning.'*

Hence, the approach followed in this study is quite different from the decomposition method. We start with the observed salt fluxes and derive the dispersion from the observed salinity distribution. This is a top-down approach as discussed in Chapter 1. In contrast, the decomposition method is a bottom-up approach that does not yield directly applicable practical equations, but which in combination with the top-down approach can yield insight into the mechanisms at work. Subsequently we may be able to attribute physical meaning to the relationships found by top-down analysis. However, there are a number of problems with the decomposition method:

1. It is done in cross sections, whereas mixing is a three-dimensional process that acts mainly in the longitudinal direction. For observations in a cross section to have significance, a large number of cross sections need to be monitored. Moreover the dominant mixing mechanism changes from one cross section to another. In one section (for instance in an ebb channel) gravitational circulation may be dominant, but in another (for instance in a cross-over point of a flood and ebb channels) it may be shear by cross-over currents and residual circulation. So an observation in one cross section does not tell us much.
2. The relative error that we make if we subtract fluxes can be very large, particularly if the residual fluxes are small compared to the momentary fluxes. In tidal hydraulics, the momentary fluxes are several orders of magnitude larger than the residual fluxes and hence the errors in the residual fluxes are often larger than the residual fluxes themselves.
3. It is highly data intensive. To determine a residual flux in a cross section one has to continue monitoring in the cross section during several tidal periods (also to account for spring–neap interaction) and sample the entire cross section at many points over the width and depth. This is both data intensive and labor intensive, and hence expensive.

It is, at the least, time consuming to investigate which of the many components is the dominant mechanism in a particular estuary under given hydrological conditions. Moreover, several scientists question the usefulness of decomposition and the correctness of linear superposition of mixing mechanisms. Important mixing processes such as the alternation between different degrees of stratification or the breaking of internal waves cannot be adequately described by the decomposition method. Rattray and Dworski (1980) state that the different components are closely interrelated, and that conclusions to be derived from this method of analysis (such as the relative importance of vertical and transverse variations to the total flux) depend on the details of the decomposition, details which are chosen by the analyst. Chatwin and Allen (1985) remark that in view of this dependency, the question of whether the transverse or the vertical dispersion is the most important salt intrusion mechanism may be less fundamental than was once believed, in that the issue is to some extent prejudged by the method of decomposition chosen.

Jay et al. (1997) observed that much research on the decomposition was done in cross sections without consideration for larger scale salt fluxes that should support the individual mechanisms. Moreover, they observed that *'the importance of the lateral terms emphasizes the three-dimensionality of estuarine transport and clearly demonstrate that two-dimensional theory cannot totally explain transport, even in narrow channelized estuaries'* (where the effect of temporal width variation due to the tide is small). Hence the three-dimensional character of mixing is crucial. One can ask oneself why researchers have lingered so long on two-dimensional (2D) analyses. There are probably two reasons. One is that much of the research to date started with laboratory flume analysis and once you are on that track, it may be difficult to explore another. The second reason may be that most of the hydraulic engineers who ventured into mixing theory started from the analysis of stratified systems, which were originally studied in 2D.

The most advanced decomposition method used is by McCarthy (1993), who integrated the 2D hydraulic and salt balance equations in an estuary with an exponential shape and found the resulting fluxes by perturbation analysis. He distinguished five components (with his terminology between quotes), 1) the landward tide-driven transport ('tidal buoyancy transport'), 2) the seaward Eulerian (non-tidal) discharge ('Eulerian buoyancy transport'), 3) the landward Stokes drift ('Stokes buoyancy transport'), 4) the landward 'diffusive buoyancy transport,' and 5) the seaward 'variable breadth diffusive transport.' The second and third components (as we have seen) are opposed and their sum equals the Lagrangean seaward salt flux. The fourth and fifth components constitute the transport driven by the salinity gradient, representing the process driven by the gravitational circulation, both vertically and laterally, but mostly laterally since the width is so much larger than the depth. Although his study did not consider the important transport mechanism by ebb–flood channel interaction, it clearly illustrated that tide-driven transport is dominant in the downstream part of estuaries, while density-driven transport is dominant at the upstream part of the salt intrusion curve. In doing so, he also emphasized the importance of the width variation and the interconnectedness of lateral, vertical, and longitudinal mixing processes. Had he included the ebb–flood channel interaction, then this would have yielded a different combination of mixing mechanisms, but still the same amount of net seaward and net landward fluxes. Hence, the inclusion of a new mechanism affects the magnitude of the others. This is also what happens in the real world. The mixing mechanisms provide feedback on each other. As a result, for the study of effective longitudinal salt intrusion it is more important to look at the mixing process as a whole than to look at each of the individual mixing mechanism.

Hence in this book, it is not the objective to further study the array of separate processes involved in dispersion. If dispersion is symbolized by an instrument, then the approach followed here is to study what the instrument does and what outer effect it has on the salt balance rather than coming to grips with all the intricacies of its inner functioning. In short, the study focuses on the apparent functioning of dispersion how dispersion counteracts the advective downstream salt transport, as a function of geometry, hydrodynamics, and density differences. Dronkers et al. (1981) supported this approach by stating that for most practical purposes, for example to determine the total salt intrusion length, it is sufficient to know the overall effect of the mixing processes rather than to understand the individual mechanisms. The analysis of the individual mixing mechanisms will help our physical understanding of how mixing works, but the addition of individual mechanisms will not yield a workable equation with which the salt intrusion can be computed and predicted in a real-life situation.

In the following, therefore, the study will focus on the effective, also called apparent, tidal average and cross-sectional average longitudinal x-dependent dispersion and how it contributes to salt intrusion. In the derivation, we shall incorporate the insight gained from analyzing the decomposition methods.

4.6 LONGITUDINAL EFFECTIVE DISPERSION

The effective dispersion incorporates all the dispersion mechanisms that counteract the advective salt transport. In a positive estuary where the salinity decreases in upstream direction, these dispersion mechanisms result in a landward transport of salt. If the dispersive salt transport is stronger than the advective salt transport, in a positive estuary, the salinity at a certain location increases with time; if it is weaker, then the salinity decreases with time. If the dispersive and advective transports are equally strong, then a tidal average state of equilibrium occurs.

In the case of equilibrium, with the salinity distribution known, the apparent tidal average dispersion coefficient D can be computed from Equation 4.8. This approach of determining the effective horizontal dispersion coefficient on the basis of the steady-state conservation of mass equation was first suggested by Stommel (1953) and later recommended by Bowden (1967). Before using this method, however, one has to check if a state of (tidal average) equilibrium indeed occurs.

Chatwin and Allen (1985) derived the following conditions under which the use of one-dimensional (1D) mixing models is justified:

1. the estuary should be long compared to the cross-sectional dimensions and the tidal excursion;
2. changes of geometry in the x-direction must be gradual.

Surely in the estuaries under study here, these conditions do not put any serious constraint on the applicability of one-dimensional mixing models. Dronkers (1982) gave an additional condition for applying 1D mixing models, the most important of which is that:

3. the time of averaging (e.g. one tidal period) should be larger than 'the timescale of cross-sectional mixing.'

This timescale is defined as the average time required for turbulent mixing in the cross section. In very wide estuaries this condition may require longer averaging times (e.g. several tidal periods), but in the Schelde, an estuary more than 10 km wide, the condition was amply met, according to Dronkers (1982). Again this condition does not pose a strict limitation on the applicability of 1D mixing models. It may be concluded that if the salt intrusion in an estuary can be adequately described by a one-dimensional tidal-averaged model, then its use is apparently justified. This is in agreement with the observation by Fischer et al. (1979) that it is better to consider the one-dimensional tidal average dispersion model as an empirical model that should be verified in practise. Fortunately, practice has shown that the domain of applicability of the 1D model in estuaries is large.

If we follow this approach then the application of the one-dimensional dispersion model depends on finding a suitable relation for D. Of the different types of relations tried in the literature, Prandle (1981) gave a good overview. He tried out the following relations:

$$D = D_0 \tag{4.19}$$

$$D \propto \frac{\partial S}{\partial x} \tag{4.20}$$

$$D \propto \left(\frac{\partial S}{\partial x}\right)^2 \tag{4.21}$$

which can be summarized as:

$$D \propto \left(\frac{\partial S}{\partial x}\right)^k \tag{4.22}$$

with $k = 0$, 1, 2, respectively.

As far as theoretical backing is concerned, the first relation ($k = 0$) occurs when the amount of energy available for mixing is uniformly distributed over the estuary. Such a situation occurs where the mixing is fully tide-driven and where both the tidal range and the tidal excursion are constant along the estuary (i.e. in an ideal estuary). The second relation ($k = 1$) corresponds with the case where the mixing is fully density-driven and where the dispersion is proportional to the moment M (see Equation 4.2) exerted by the two hydrostatic forces shown in Figure 2.2. Hence, the density-driven dispersion is proportional to the salinity gradient. If one assumes that the density-driven dispersion works laterally rather than vertically, described by Equation 4.3 proposed by Fischer et al. (1979), then one arrives at $k = 2$.

With regard to estuary shape, Prandle (1981) obtained some unexpected results. He showed that in flumes and in estuaries with almost constant cross section

(Rotterdam Waterway) Equations 4.20 and 4.21 performed best, but that in estuaries with a pronounced funnel shape, such as the Thames, the St. Lawrence, the Delaware, and the Bristol Channel, very good results were obtained with the simple Equation 4.19.

What is surprising is that some suggested the opposite. Several researchers (among others West and Broyd, 1981) mentioned that density-driven dispersion ($k = 1$) prevails in deep and wide estuaries and tide-driven dispersion ($k = 0$) in narrow and shallow estuaries. This appears in contradiction with what we found earlier that estuaries with a pronounced funnel shape have dome-shaped salt intrusion curves, whereas estuaries with almost a constant cross section have recession type intrusion curves. Near the mouth of an estuary with a dome-type intrusion curve (being an estuary with a pronounced funnel shape where the mixing process is claimed to be density-driven) hardly any density gradient occurs, and hence hardly any density-driven mixing can occur. One reason for this paradox is that the researchers who studied deep and wide estuaries with dome-shaped intrusion curves and who concluded that the mixing was primarily density-driven, carried out their investigations somewhere in the middle reach of the salt intrusion curve. So, although they were correct in identifying density-driven mixing as the predominant mechanism in the middle reach of wide estuaries, hardly any density-driven mixing occurred in the wider part, near the mouth, of these estuaries.

So it is more correct to draw another conclusion. A narrow and prismatic estuary like the Rotterdam Waterway has a high level of stratification and a rather steep (recession-type) salt intrusion curve. The dominant mixing mechanism is gravitational circulation, which is best described by Equations 4.20 or 4.21. A wide funnel-shaped estuary, such as the Thames, the Delaware, or the Schelde, has a dome-shaped intrusion curve and a small density gradient near the mouth. In the downstream part of such an estuary, the dominant mechanism is tide-driven mixing in the form of ebb and flood channel interaction (as long as the estuary is wide enough for separate ebb and flood channels to develop: $B/h > 100$). In the reach where the dome-shaped intrusion curves bends down, and hence the salinity gradient is strong, gravitational circulation becomes dominant with a strong lateral component.

As a result, the dispersion is the highest near the estuary mouth, decreasing in upstream direction. Because the gravitational circulation is proportional to the density gradient, the dispersion reduces with the salinity gradient until it becomes very small near the toe of the intrusion curve. Beyond the toe of the intrusion curve the dispersion is dominated by turbulent diffusion, which is small compared to the other mechanisms.

Many researchers, such as Preddy (1954), Kent (1958), Ippen and Harleman (1961), and Stigter and Siemons (1967), recognized that the effective dispersion is the highest near the estuary mouth and that it decreases upstream to become zero, or virtually zero, near the toe of the salt intrusion curve. The challenge now is to find a relationship that applies well in all estuaries, whether funnel-shaped or of almost uniform cross section, which is easy to incorporate in a predictive model and which is physically appealing. More specifically:

a. the relation should be dimensionally sound, with dimensionless coefficients,
b. the dispersion should decrease in upstream direction,
c. the dispersion should be large near the mouth of funnel-shaped estuaries,
d. the relation should be continuous and easy to apply.

The following relation suggested by Savenije (1986, 1989, 1992a, 1993a,b) performs well against these criteria:

$$\frac{D}{D_0} = \left(\frac{S}{S_0}\right)^K \tag{4.23}$$

where K is the dimensionless Van den Burgh's coefficient (Van den Burgh, 1972) which lies between zero and unity. This relationship has been widely tested and successfully applied in about 20 estuaries worldwide. It is interesting to see that in contrast to Prandle's relations, summarized by Equation 4.22, in this equation the salinity is used instead of the salinity gradient.

If we use this simple relationship, then a dome-shaped intrusion curve (which occurs in wide and deep estuaries) results in a high value of the dispersion coefficient near the mouth, decreasing slowly in upstream direction to become zero at the toe of the intrusion curve. In a narrow estuary, where the salinity decreases steeply, the salinity curve has an exponential decline (recession shape). In an exponential function, the function value is directly proportional to its gradient. So $\mathrm{d}S/\mathrm{d}x$ is proportional to S, and the dispersion is proportional to the salinity gradient to the power K. This is in agreement with the theory. So, Equation 4.23 applies both to tidal-driven dispersion (particularly ebb–flood channel exchange) and to gravitational circulation. This equation describes the mixing process well: near the mouth, in the middle, and near the toe, and both in funnel-shaped estuaries and in estuaries with a long convergence length. The coefficient K is obtained through calibration.

Figure 4.6 illustrates how Equation 4.23 performs in relation to an observed salt intrusion curve in the Pungué estuary with a dome-shaped intrusion curve. The thick line is a dimensionless salt intrusion curve S/S_0 that fits the observations made on 3 October 1993. The observations at HWS have been translated half a tidal excursion to the left and the observations at LWS half a tidal excursion to the right. The dimensionless dispersion curve has been computed with the steady-state salt balance equation (Equation 4.8), but it also completely corresponds with Equation 4.23. For the case where $K = 1$, the curves of D/D_0 and S/S_0 coincide. So the line indicated by D/D_0 is the line that fits the observed effective 1D tidal-average dispersion.

Finally, the dashed line corresponds with Equation 4.20, and reflects the density-driven dispersion. It is not possible to scale the density-driven mixing exactly;

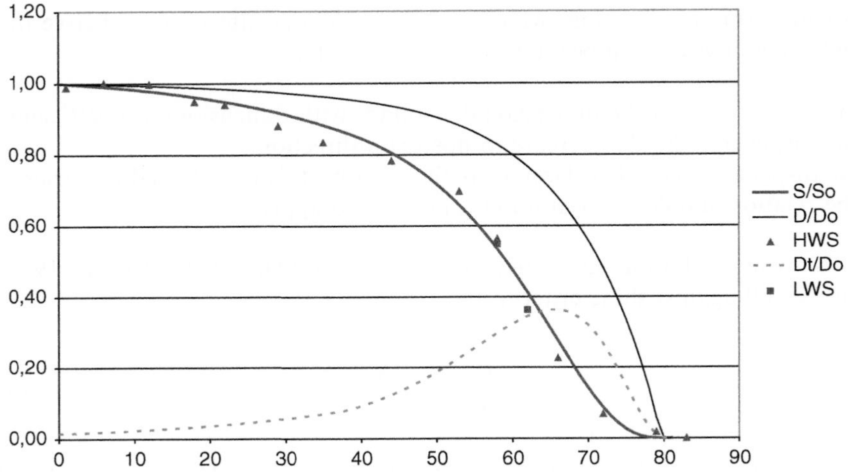

Figure 4.6 Dispersion equation derived from the salt intrusion curve measured in the Pungué estuary on 3/10/1993.

the position of the curve is merely indicative. If we assume that the total dispersion is the sum of the tide-driven dispersion and the density-driven dispersion, then the difference between these two curves equals the tide-driven dispersion. It can be seen in Figure 4.6 that the tide-driven dispersion near the mouth is very large due to ebb–flood channel interaction and reduces as the estuary becomes narrower. After the point where the salinity gradient is at its steepest the density-driven mixing becomes more important. The value of K obtained in this estuary is 0.3, suggesting that tide-driven mixing is more important than density-driven mixing.

Figure 4.7 shows a similar graph for the Maputo estuary with a bell-shaped intrusion curve. It is based on observations made on 29 May 1984. Here, we also see that tide-driven mixing is dominant in the wider part of the estuary, but less so as in the strongly funnel-shaped Pungué estuary. Upstream from the point where the salinity gradient has its maximum, the density-driven mixing is dominant. The Maputo has a trumpet shape with a short convergence length near the mouth (of 3.5 km) and a longer convergence length upstream (16 km). The inflection point of the density-driven mixing lies exactly at the point where the change of the convergence length takes place: the upper reach having a more prismatic character and the lower reach a pronounced funnel shape. The value of K in this estuary is also 0.3, with tide-driven mixing being more important than density-driven mixing.

Figure 4.8, finally, presents an estuary with a near prismatic channel: The Limpopo, with a convergence length of 130 km in the upstream reach (and 50 km in the downstream reach). As a result the intrusion curve has a recession shape, which is an indication of gravitational circulation being the dominant mechanism. We see that this is indeed the case. Consequently, we expect a higher value of K which is indeed the case K being equal to 0.5.

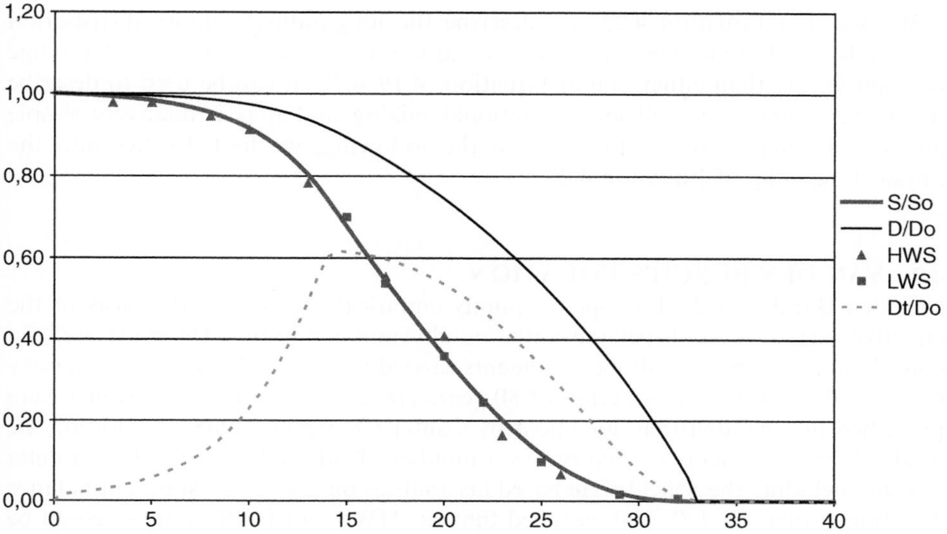

Figure 4.7 Dispersion equation derived from the salt intrusion curve measured in the Maputo estuary on 29/5/1984.

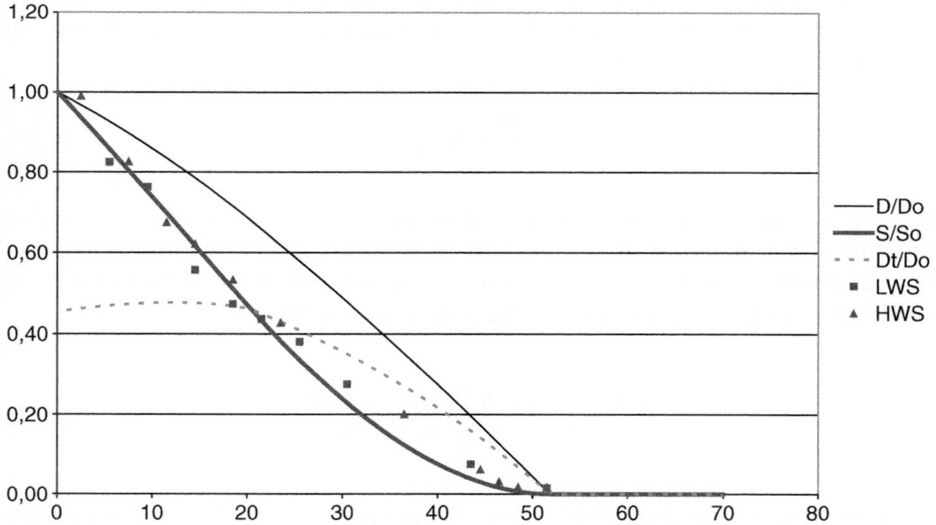

Figure 4.8 Dispersion equation derived from the salt intrusion curve measured in the Limpopo estuary on 24/7/1994.

We see that Equation 4.23 can describe the longitudinal salinity distribution very well in different types of estuaries. Equation 4.23 has a much wider range of applicability than either one of Equations 4.19–4.22. It can be used to describe tide-driven mixing as well as gravitational mixing and it is a relatively simple and dimensionally correct formula. In the following, we look further into the physical meaning of Equation 4.23.

4.7 VAN DEN BURGH'S EQUATION

Van den Burgh (1972) developed a purely empirical method on the basis of the effective tidal average dispersion under equilibrium conditions. He made use of a considerable number of salt measurements carried out in the Rotterdam Waterway (excavated in 1868) over a period of 80 years (1892–1971), the first of which were published in a handwritten document by Canter Cremers in 1905. In addition, he used salt measurements carried out in a number of tidal inlets of the Dutch delta system, including the Schelde. He based his analysis on the steady-state salt balance equation (Equation 4.8), and assumed that the HWS and LWS salinities could be obtained through longitudinal translation over half the tidal excursion, which indeed is correct.

Analyzing the salt distributions surveyed in the Rotterdam Waterway over the period 1905–1971, he observed that the effective dispersion obtained from Equation 4.8 decreased in upstream direction. Plotting the longitudinal variation of the effective dispersion against the velocity of the fresh water discharge, he found a direct proportionality:

$$\frac{\partial D}{\partial x} = KU_{\mathrm{f}} \tag{4.24}$$

where K is Van den Burgh's dimensionless coefficient. Since U_{f} has a negative sign, the dispersion decreases in upstream direction. This relation, known as Van den Burgh's equation, can be demonstrated to be the same as Equation 4.23.

Differentiation of Equation 4.23 with respect to x yields:

$$\frac{\partial D}{\partial x} = K\frac{D_0}{S}\left(\frac{S}{S_0}\right)^K \frac{\partial S}{\partial x} = K\frac{D}{S}\frac{\partial S}{\partial x} \tag{4.25}$$

It can be directly seen that combination of this equation with the steady-state salt balance equation, Equation 4.8, yields Equation 4.24.

A major difficulty Van den Burgh had in determining the effective dispersion was that the exact amount of fresh water discharge in the estuary was not always well known. Although a good estimate of the Rhine discharge on entering The Netherlands (at Lobith) was available, the division of water over the many branches was difficult to determine at that time. In addition, the time required for the salinity distribution in the estuary to adjust to the flow recorded at Lobith was not known. Van den Burgh used the results of an electro-analogous model of the Rhine delta to obtain the fresh flows in the estuary. However, the inaccuracy in the methodology was probably still great, which makes it the more surprising that he came up with this important finding.

Van den Burgh named his approach the 'advective method.' His method has the advantage that, for its derivation, it does not require a constant cross section. Moreover, the relation between $\partial D/\partial x$ and U_f is dimensionally sound.

4.7.1 The physical meaning of Van den Burgh's *K*

It can be shown mathematically that Van den Burgh's coefficient is a sort of 'shape factor' influencing the shape of the salt intrusion curve. Kranenburg, in a personal communication, indicated this to the author for a channel of constant cross section. The application of this approach to a channel with an exponentially varying cross section is explored in the following.

Differentiation of Equation 4.8 with respect to x yields:

$$QS' = D'A'S + D'AS' + DAS'' \tag{4.26}$$

where a single accent indicates the first partial derivative and a double accent the second partial derivative with respect to x.

Since $A'/A = -1/a$ and $D = (U_f S/S')$, Equation 4.26 can be elaborated into:

$$D' = U_f - \frac{DA'}{A} - \frac{DS''}{S'} = U_f\left(1 + \frac{S}{aS'} - \frac{SS''}{(S')^2}\right) \tag{4.27}$$

Substitution of Van den Burgh's equation yields:

$$K = 1 + \frac{S}{aS'} - \frac{SS''}{(S')^2} \tag{4.28}$$

The influence of K and a on the shape of the salt intrusion curves can be made clear by scaling. We scale S by the sea salinity to obtain the dimensionless salinity:

$$\varsigma(\xi) = \frac{S}{S_0} \tag{4.29}$$

where $\xi = x/L$. Elaboration of Equation 4.28 then leads to:

$$\frac{\varsigma\varsigma''}{(\varsigma')^2} = (1 - K) + \frac{\varsigma}{\varsigma'}\frac{L}{a} \tag{4.30}$$

where ς' is the first derivative of ς with respect to ξ, and ς'' is the second derivative. The left-hand side is the shape function which is influenced by two terms on the right-hand side. It can be easily seen that the shape function is positive if the curvature ς'' is positive (because $\varsigma \geq 0$ for all ξ on [0,1]).

In the integration of the steady-state salt balance equation the boundary condition used is that $\varsigma' = 0$ where $\varsigma = 0$, at the upstream end of the intrusion curve. Since, in a positive estuary, the gradient of the salinity is negative for all ξ on [0,1], ς' can only become zero at the toe of the intrusion curve if somewhere within this interval the curvature ς'', and hence the shape function, becomes positive. Because the second term of the right-hand side of Equation 4.30 is always negative in a positive estuary (since $\varsigma' < 0$), the curvature can only become positive in the interval [0,1] if the first term on the right-hand side is positive; this is the case when $K < 1$. Hence there is an upper limit to the value of K. Since the lower boundary of K is zero, it follows that $0 < K < 1$.

Dome-shaped (type 3) intrusion curves have a negative curvature ($\varsigma'' < 0$), at least in the downstream part of the estuary, and recession-shaped (type 1) intrusion curves have a positive curvature ($\varsigma'' > 0$). The bell-shaped (type 2) intrusion curve is a mixture of these two, dome-shaped near the mouth of the estuary and recession-shaped at the toe of the intrusion curve. Hence the left-hand side of Equation 4.30 is negative in type 3 estuaries, and positive in type 1 estuaries. Therefore, dome-shaped intrusion curves occur when the term containing K is small with respect to the absolute value of the term containing L/a (meaning a large value of K and a large value of L/a) and recession-shaped intrusion curves in the opposite case (K is small and L/a is small). This is interesting since it implies that K and L/a are shape factors influencing the shape of the salt intrusion curves.

4.7.2 Correspondence with other methods

Van den Burgh's method has similarities with a number of methods developed by other authors. Preddy (1954), Kent (1958), Ippen and Harleman (1961), and Stigter and Siemons (1967) already recognized that the effective dispersion decreased in upstream direction. McCarthy (1993) clearly demonstrated how the dispersion decreases in upstream direction, to become zero near the toe of the salt intrusion curve.

The theory demonstrating the largest similarity with Van den Burgh's method is that of Hansen and Rattray (1965). They limited their theory to the central zone of a narrow estuary of constant cross section. In addition, they assumed that the salinity in the central zone would decrease linearly (no curvature: $\varsigma'' = 0$) in upstream direction. Both are very strong limitations which Van den Burgh's method does not have. On the basis of these strong assumptions, they arrived at three so-called similarity conditions for the two-dimensional vertical description of velocity and salinity, being: that the vertical turbulent viscosity and dispersion were constant along the estuary and that the horizontal diffusive (= tide-driven) dispersion D_t was defined by:

$$\frac{\partial D_t}{\partial x} = U_f \tag{4.31}$$

which is indeed what we would obtained from Equaiton 4.27 by substitution of $a \to \infty$ (constant cross section) and $\varsigma'' = 0$ (no curvature).

Moreover, they defined the parameter ν by: the fraction of the salt advected seaward with the river discharge ($\nu U_f S$) that is balanced by the upstream salt flux associated with tidal dispersion ($D_t\, \partial S/\partial x$). Consequently, by using the steady-state conservation of mass equation for salt it can be seen that, ν equals the ratio of the tide-driven dispersion D_t to the effective (=total) dispersion D:

$$\nu = \frac{D_t}{D} \tag{4.32}$$

Fischer et al. (1979) gave a wider definition of ν as the fraction of the total landward transport of salinity caused by all dispersion mechanisms other than the density-driven circulation, meaning that D_t incorporates all dispersion mechanisms other than density-driven dispersion. If $\nu = 0$ then the dispersion is fully density-driven; if $\nu = 1$, the dispersion is driven by other mechanisms. Thus combination of Equations 4.31 and 4.32 leads to:

$$\frac{\partial D}{\partial x} = \frac{1}{\nu} U_f \tag{4.33}$$

which is identical to Van den Burgh's equation for $K = 1/\nu$. Apparently, in an estuary with constant cross section and with a salinity distribution without curvature, Van den Burgh's $K = 1/\nu$. In Hansen and Rattray's definition, Van den Burgh's coefficient is the proportion of total effective dispersion to the tide-driven dispersion. If $K = 1$, the dispersion is fully tide driven. If $K > 1$, the influence of density-driven dispersion becomes more pronounced. This is in conflict with the limits to K earlier found, but one should realize that Hansen and Rattray's relationship was derived under strongly limiting assumptions such as a constant cross-sectional area and $\varsigma'' = 0$. In an exponentially shaped estuary or in an estuary with a non-linear salinity distribution, Hansen and Rattray's result would have been different, and K would not necessarily have to be larger than unity.

It has been observed earlier that each estuary has a salt intrusion curve with a characteristic shape. This indicates that K is closely related to the general characteristics of an estuary such as its shape and average hydraulic conditions.

4.8 GENERAL EQUATION FOR LONGITUDINAL DISPERSION

Integration of Van den Burgh's equation taking into account the exponential variation of the cross-sectional area yields the following equation:

$$\frac{D}{D_0} = 1 - \beta\left(\exp\left(\frac{x}{a}\right) - 1\right) \tag{4.34}$$

where:

$$\beta = -\frac{KaQ_f}{D_0 A_0} \tag{4.35}$$

where D_0 is the dispersion at the estuary mouth, A_0 is the cross-sectional area at the estuary mouth, a is the convergence length, and Q_f is the fresh water discharge (which is negative). This is a simple equation relating the longitudinal effective dispersion to the longitudinal ordinate, the estuary geometry, and the river discharge. We can see that estuary shape is prominently present through both a and A_0. For the equation to become predictive, an expression for D_0 and K needs to be found. This is done in the following chapter.

5

Salt intrusion in alluvial estuaries

There are different types of salt intrusion, depending on topography, hydrology, and tide. A classification is presented with the related characteristics and conditions for occurrence.

The salt intrusion in alluvial estuaries is generally of the well-mixed or partially mixed type, particularly in the period when it matters: the dry season. This salt intrusion can be adequately described by the one-dimensional dispersion equation. When this equation is combined with equations describing topography and dispersion, it yields a predictive equation that can be used for both unsteady and steady state, particularly for the tidal average situation, high water slack, and low water slack. The influence of rainfall and evaporation on salt intrusion is also taken into account, which sometimes is crucial to understand the salinity distribution. Under extreme circumstances hypersalinity can occur, which is illustrated with case material.

5.1 TYPES OF SALT INTRUSION AND SHAPES OF SALT INTRUSION CURVES

In an estuary of the well-mixed type, the variation of the salinity along the longitudinal axis of the estuary is gradual. This implies that if a continuous survey is done along an estuary, e.g. by a moving boat or by simultaneous observations along the estuary, a smooth curve can be fitted through the observed cross-sectional averaged salinities. The shape of the curve, however, can differ widely depending on the situation at hand.

A number of designations will be used to characterize salt intrusion curves of a particular shape. It is a classification to help identify certain types of salt intrusion, which, as will be shown further on, have certain relations to both the geometric shape of an estuary and the hydrology. The following types are distinguished (see Figure 5.1):

- type 1, recession shape
- type 2, bell shape
- type 3, dome shape
- type 4, humpback shape

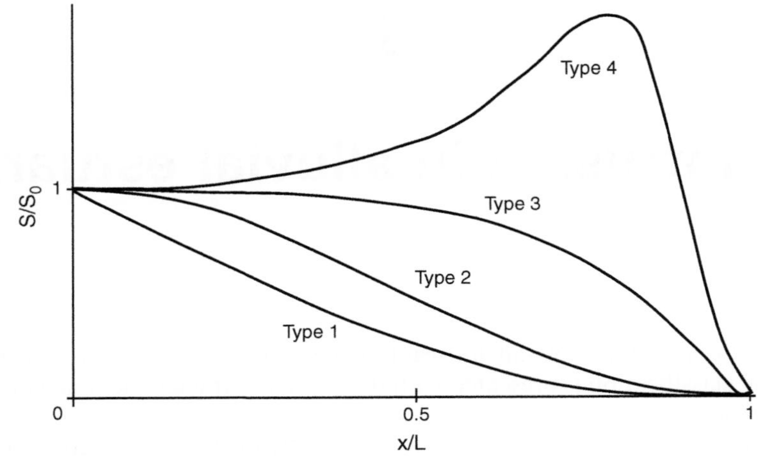

Figure 5.1 Four types of salt intrusion curves.

Type 1 is an intrusion curve with a logarithmic convex shape; the salinity gradient at the estuary mouth is steep. Examples of this type are (see Table 2.2) the Chao Phya in Thailand and the Limpopo in Mozambique, which are close to prismatic, straight, and narrow estuaries (a large value of b). Type 3 is completely the opposite: it has a concave shape and the salinity gradient at the mouth is small. Examples of this type are: the Delaware in the USA, the Thames in the UK, and the Schelde in the Netherlands. These are all wide channels with a pronounced funnel shape (a small value of b). Type 2 is not a transition from type 1 to 3, it is rather a mixture of the two. It starts concave, but within 50 percent of the salt intrusion length, it changes into a convex shape. Examples are the Maputo in Mozambique and the Mae Klong in Thailand. These estuaries are close to prismatic upstream (a large value of b), but strongly funnel-shaped near the mouth (a small value of b).

Apparently, these three types of intrusion curves are very much linked to the geometry of the estuary. All estuaries studied appear to have a fixed type of curve shape irrespective of the hydrological conditions (within the hydrological limits between which well-mixed salt intrusion occurs): the intrusion may increase or decrease, the shape type remains unchanged.

The fourth type is an exception in this respect. A humpback shape is entirely the result of a rainfall deficit or an evaporation excess. Evaporation can change a bell-shaped intrusion into a dome-shaped intrusion and eventually into a hypersaline intrusion. Generally, in well-mixed estuaries, the salinity reduces in upstream direction. Such an estuary is a normal or 'positive' estuary (see Section 1.2). In a positive estuary, the sum of the fresh water inflow and the direct rainfall on the estuary surface exceeds the evaporation (Dyer, 1973). In some arid parts of the world, however, so-called hypersaline or 'negative' estuaries occur where evaporation exceeds the sum of rainfall and runoff. In a hypersaline estuary, the salinity increases in the upstream direction until it reaches a maximum after which it decreases depending on the amount of fresh water inflow. The maximum salinity to be reached is saturation level (363 kg/m^3 at 20°C). Such hypersaline estuaries occur for example in the Sahel: the Saloum and the Casamance (Pagès and Citeau, 1990), and in some tropical North Australian estuaries (Wolanski, 1986). Although in all estuaries there is, in principle, an influence of direct rainfall and evaporation on the salinity of the estuary water, this effect is not important in most of the estuaries. In Sections 5.3 and 5.7, more attention is paid to this phenomenon.

This classification of salt intrusion curves is a purely descriptive one. In the following sections the salt intrusion will be dealt in an analytical way making use of the equations of conservation of mass for water and salt.

5.2 SALT BALANCE EQUATIONS

In analogy with the 1-D mass balance equation for water, the mass balance equation for dissolved salts states that the sum of the change in salt load over time and the change of the salt fluxes over the distance should be zero, or in case of a source of salts, equal to the source. Hence the cross-sectional averaged salt balance equation reads:

$$r_S \frac{\partial As}{\partial t} + \frac{\partial F}{\partial x} = 0 \quad (5.1)$$

where:

- r_S is the storage width ratio defined earlier in Chapter 2
- $F = F(x,t)$ is the mass flux of salt in kg/s averaged over the cross-sectional area A;
- $s = s(x,t)$ is the salinity in kg/m^3.

In contrast to the water balance, Equation 2.2, the salt balance equation does not have a source term. In Equation 5.1, the source term can be disregarded, unless salt is picked up from the bottom or from marginal salt flats. Salt deposition by rainfall is marginal and evaporation does not carry any salt either, so for all practical purposes in alluvial estuaries the source term may be disregarded.

Subsequently, the mass flux is defined as:

$$F = \int\int Us \mathrm{d}z \mathrm{d}y \quad (5.2)$$

where $U = U(x,y,z,t)$ and $s = s(x,y,z,t)$ are the water velocity and salt concentration at a certain point (y,z) in the cross section. Equation 5.1 can be elaborated by making use of Equation 2.2, the continuity equation of water, into:

$$r_S A \frac{\partial s}{\partial t} - s\frac{\partial Q}{\partial x} + \frac{\partial F}{\partial x} = -sR_s \tag{5.3}$$

where Q is the discharge. We see that through the water balance, the source term enters into the equation.

The mass flux F is generally decomposed into an advective and a dispersive term:

$$F = Qs - AD\frac{\partial s}{\partial x} \tag{5.4}$$

where $D = D(x,t)$ is the longitudinal dispersion coefficient in m^2/s. Differentiation of Equation 5.4 leads to:

$$\frac{\partial F}{\partial x} = Q\frac{\partial s}{\partial x} + s\frac{\partial Q}{\partial x} - \frac{\partial}{\partial x}\left(AD\frac{\partial s}{\partial x}\right) \tag{5.5}$$

Combination of Equations 5.3 and 5.5 leads to:

$$r_S A \frac{\partial s}{\partial t} + Q\frac{\partial s}{\partial x} - \frac{\partial}{\partial x}\left(AD\frac{\partial s}{\partial x}\right) = -sR_s \tag{5.6}$$

The storage width ratio in Equation 5.6 is often disregarded. This is not correct, but in steady state models, the effect of the storage width is obviously not present. In dynamic models where it is disregarded, it can be compensated in the first term by increasing the value of A_0 (assuming a larger cross-sectional area at the mouth), but obviously, that would require an adjustment of the dispersion as well to correct the error thus introduced in the third term.

Equation 5.6 is the unsteady state one-dimensional salt balance equation. Separation of the discharge Q into a tidal component and a fresh water component yields:

$$r_S A \frac{\partial s}{\partial t} + (Q_t + Q_f)\frac{\partial s}{\partial x} - \frac{\partial}{\partial x}\left(AD\frac{\partial s}{\partial x}\right) = -sR_s \tag{5.7}$$

where $Q_t = Q_t(x,t)$ is the tidal discharge, which has a time-average value of zero, and $Q_f = Q_f(t)$ is the fresh water discharge of the river(s) entering the estuary. The time scale of the temporal variation of the two discharges is different. The tidal discharge variation has a time scale of hours, whereas the fresh water discharge variation has a time scale of days to months. It should be noted that, since the positive x-direction has been chosen in upstream direction, Q_f has a negative value. In the remainder of this section, the source term will be neglected. In Section 5.3, the effect of rainfall and evaporation on this term will be dealt with.

Situation at High Water Slack and Low Water Slack

A special case of Equation 5.7 occurs at high water slack (HWS), when the direction of flow changes from upstream to downstream. In case of a mixed-type wave (see Section 2.2) the time at which HWS occurs—some time after reaching high water (HW)—grows later as the tidal wave moves upstream. Hence at each point along the estuary, HWS occurs at a different time. At HWS—by definition—the tidal discharge Q_t is zero.

As was done for the discharge, the rate of change of the salinity $\partial s/\partial t$ can also be decomposed into a tidal component—with a periodicity equal to the tidal period—and a long-term component. Since in a situation where the fresh water discharge is constant, the maximum salinity is reached when Q_t is zero at HWS, it is reasonable to assume that also when Q_f is not constant, the tidal component of $\partial s/\partial t$ is zero at HWS. Hence Equation 5.7 can be modified for HWS into:

$$r_S A \frac{\partial s_{HWS}}{\partial t} + Q_f \frac{\partial s_{HWS}}{\partial x} - \frac{\partial}{\partial x}\left(A_{HWS} D_{HWS} \frac{\partial s_{HWS}}{\partial x}\right) = 0 \tag{5.8}$$

where $\partial s_{HWS}/\partial t$ is the long-term variation of the salinity at HWS. If the long-term variation is negligible (Q_f is constant), a situation of equilibrium occurs in which there is a balance between the second and the third terms. In that case, Equation 5.8 can be integrated with respect to x. The uppercase $S = S(x)$ is used to indicate the steady state salinity. Since the fresh water discharge Q_f may be considered invariant with x, integration under the boundary condition that $S = S_f$ (the fresh water salinity) and $\partial S/\partial x = 0$ when $x \to \infty$, yields:

$$Q_f(S_{HWS} - S_f) - A_{HWS} D_{HWS} \frac{\partial S_{HWS}}{\partial x} = 0 \tag{5.9}$$

An analogous derivation can be made for low water slack (LWS):

$$Q_f(S_{LWS} - S_f) - A_{LWS} D_{LWS} \frac{\partial S_{LWS}}{\partial x} = 0 \tag{5.10}$$

The subscript LWS refers to the situation at LWS.

The curves described by Equations 5.9 and 5.10 represent the two sets of points in the S-x plane that occur at HWS and LWS respectively. Since HWS and LWS represent the upper and lower extremes of the salt intrusion, they form two envelope curves between which the salinity varies (see Figure 5.2). Instantaneous salt intrusion curves fall within the two envelopes.

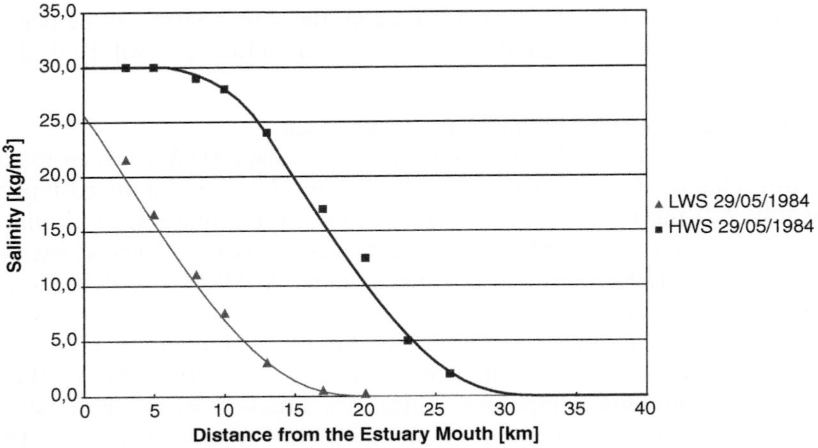

Figure 5.2 Envelope curves of salinity intrusion at High Water Slack (HWS) and Low Water Slack (LWS). The example is taken from the Maputo estuary. The observations made on 29/5/1984 are indicated by symbols, the envelope curves by drawn lines.

Mean tidal situation

Another special case is the tidal average situation, which follows from averaging over a tidal cycle under the following first order approximations:

$$\frac{1}{T}\int_0^T A\frac{\partial s}{\partial t}\mathrm{d}t \approx A_{\mathrm{TA}}\frac{\partial s_{\mathrm{TA}}}{\partial t} \tag{5.11}$$

$$\frac{1}{T}\int_0^T Q_{\mathrm{t}}\frac{\partial s}{\partial x}\mathrm{d}t \approx 0 \tag{5.12}$$

$$\frac{1}{T}\int_0^T Q_{\mathrm{f}}\frac{\partial s}{\partial x}\mathrm{d}t \approx Q_{\mathrm{f}}\frac{\partial s_{\mathrm{TA}}}{\partial x} \tag{5.13}$$

$$\frac{1}{T}\int_0^T \frac{\partial}{\partial x}\left(AD\frac{\partial s}{\partial x}\right)\mathrm{d}t \approx \frac{\partial}{\partial x}\left(A_{\mathrm{TA}}D_{\mathrm{TA}}\frac{\partial s_{\mathrm{TA}}}{\partial x}\right) \tag{5.14}$$

where the subscript TA refers to the tidal average (TA) situation, and D_{TA} is the tidal and cross-sectional average, x-dependent, dispersion. In particular, the latter three assumptions are inaccurate near the toe of the salt intrusion curve, where $\partial s/\partial x$ may vary strongly with time. This led Fischer et al. (1979, p. 270) to remark that efforts made to derive a time-averaged dispersion equation starting from Equation 5.7 have not been wholly satisfactory because of the assumptions required. Making use of the above assumptions, Equation 5.7 can be modified into:

$$r_S A_{TA} \frac{\partial s_{TA}}{\partial t} + Q_f \frac{\partial s_{TA}}{\partial x} - \frac{\partial}{\partial x}\left(A_{TA} D_{TA} \frac{\partial s_{TA}}{\partial x} \right) = 0 \tag{5.15}$$

Fischer et al. (1979) remark that in view of the weakness of the assumptions, it is better to consider Equation 5.15 as an empirical model that should be verified in practice. However, the fact that Equation 5.15 has been used successfully in many mathematical models appears to be sufficient reason to adopt this model.

In support of this, O'Kane (1980) demonstrated that Equation 5.15 follows from Equation 5.7 in an oscillating framework, where the observer maintains a constant volume of water between himself and the head of the estuary. The only assumption required in his derivation is that the dispersion is devoid of harmonics. In fact, in a one-dimensional model, O'Kane's framework is Lagrangean as long as the amount of fresh water inflow is negligible compared to the tidal volume. In well-mixed estuaries, this is an acceptable assumption. O'Kane's equation can be applied equally well to the tidal average situation, or to slack; the difference between HWS, LWS, or TA being merely a longitudinal translation (see Figure 5.2). In all three cases, the rate of change $\partial s/\partial t$ reflects the long-term variation of the salinity. In this respect, the dispersion D_{TA} should be considered as an 'effective' tidal average dispersion, since it is a bulk parameter that expresses the result of all mixing processes that occur within a tidal cycle.

Although Equation 5.15 may be considered correct, in this book, we shall not focus on the TA situation. In our approach, the HWS and the LWS situations are considered more relevant, providing the extremes between which salt intrusion is bound. Particularly, the HWS situation is important since it determines the maximum salt intrusion that occurs during a given day. For design purposes, this is the critical situation.

In the special case where equilibrium occurs between the advective and dispersive terms: $\partial s_{TA}/\partial t = 0$ and steady state occurs. In that case Equation 5.15 can be integrated with respect to x to yield:

$$Q_f(S_{TA} - S_f) - A_{TA} D_{TA} \frac{\partial S_{TA}}{\partial x} = 0 \tag{5.16}$$

where S_{TA} represents the mean tidal steady-state salinity. The boundary condition used is that $S_{TA} = S_f$ and $\partial S_{TA}/\partial x = 0$ when $x \rightarrow \infty$. The river water salinity S_f is generally small as compared to the salinity in the estuary ($S_{TA} \ll S_f$) and is often disregarded.

Although some researchers (Ippen, 1966b; Hansen and Rattray, 1965) have done so, it is not necessary to assume that A_{TA} and D_{TA} be constant with x. Hence, both A_{TA} and D_{TA} are still functions of x.

Similarity between HWS, LWS, and TA situation

There is a large similarity between Equations 5.9, 5.10, and 5.16. All three equations, in the particular situation for which they apply (HWS, LWS, or TA) are of the form:

$$S_i - S_f = c_i \frac{\partial S_i}{\partial x} \tag{5.17}$$

where $i = 1,2,3$ indicates the three different states: HWS, LWS, and TA, and where c_i is an x-dependent coefficient equal to the ratio between the dispersion coefficient and the fresh water velocity, which is different for each state. Hence, the difference between the three equations lies in c_i and in the different downstream boundary conditions belonging to each state. It will be shown in Section 5.5 that, as a result of the expression used for c_i, the three curves have identical shapes but different positions along the x-axis.

5.3 INFLUENCE OF RAINFALL AND EVAPORATION

The direct rainfall on an estuary and the evaporation from an estuary are described mathematically by the same parameter r, representing net rainfall in m/s (the balance of rainfall and evaporation). If r is negative, then evaporation predominates.

The continuity equation for water, Equation 2.2 changes if rainfall is added to the water balance:

$$r_S \frac{\partial A}{\partial t} = -\frac{\partial Q}{\partial x} + Br \tag{5.18}$$

where B is the estuary width. A combination of Equation 5.18 with Equation 5.1 and comparison with Equation 5.3 yields an expression for the source term, as an effect of rainfall or evaporation:

$$r_S A \frac{\partial s}{\partial t} - s\frac{\partial Q}{\partial x} + \frac{\partial F}{\partial x} = -sBr \tag{5.19}$$

The source term has a negative sign. It is negative if rainfall is added to the system, leading to a dilution of saline water, and positive in case of evaporation, leading to a higher concentration of salt.

To simulate the effect of rainfall and evaporation, the mean tidal salt balance equation, i.e. Equation 5.15 is used, to which the source term is added:

$$r_S A\frac{\partial s}{\partial t} + Q_f\frac{\partial s}{\partial x} - \frac{\partial}{\partial x}\left(AD\frac{\partial s}{\partial x}\right) = -sBr \tag{5.20}$$

Rainfall and evaporation affect the salt balance in two ways. One is through the source term, another is through the fresh discharge Q_f.

Integration of Equation 5.18 with respect to x for the mean tidal situation where $Q = Q_f$ and under the assumption that the long-term variation in the cross-sectional area $\partial A/\partial t$ is negligible, leads to:

$$\int_x^\infty \mathrm{d}Q_f = \int_x^\infty Br\mathrm{d}x \tag{5.21}$$

The integration is from x to infinity, being the length over which the tidal influence is felt, which is a length several times larger than the width convergence length b. There, the fresh water discharge equals the river discharge Q_r which is supposed to be measured outside the tidal influence. Elaboration of Equation 5.21 yields:

$$Q_f(x) = Q_r - Brb \tag{5.22}$$

It means that if r is positive, $Q_f(x)$ grows more negative (Q_r is negative since the positive x-direction points upstream) towards the mouth of the estuary. If r is negative, then there is a point where Q_f becomes zero, after which it changes into a positive direction, attracting water from the sea.

The salt balance equation hence modifies into:

$$r_S A\frac{\partial s}{\partial t} + (Q_r - Brb)\frac{\partial s}{\partial x} - \frac{\partial}{\partial x}\left(AD\frac{\partial s}{\partial x}\right) = -sBr \tag{5.23}$$

Depending on the relative size of each term in Equation 5.23, the effect of rainfall and evaporation is important or small. The relative size of each term depends on both hydrological and geometrical parameters, which can vary strongly from estuary to estuary. In most estuaries, however, the importance is small, which is one of the reasons that not much attention is given to this issue in the literature. Savenije (1988) showed that evaporation is important to describe the salinity distribution in the Gambia. Savenije and Pagés (1992) showed that this was also the case for the estuaries north and south of the Gambia: the Saloum and the Casamance, leading to hypersaline conditions.

Scaling of the salt balance equation

To analyze the relative size of the terms in Equation 5.23, the terms of the equation are scaled. Elaboration yields:

$$r_S A \frac{\partial s}{\partial t} + (Q_r - Brb)\frac{\partial s}{\partial x} - \frac{\partial A}{\partial x} D \frac{\partial s}{\partial x} - \frac{\partial D}{\partial x} A \frac{\partial s}{\partial x} - AD\frac{\partial^2 s}{\partial x^2} = -sBr \tag{5.24}$$

Using the exponentially varying cross-sectional area of Equation 2.38, Equation 5.24 can be further elaborated:

$$r_S \frac{\partial s}{\partial t} + \frac{Q_r}{A}\frac{\partial s}{\partial x} - \frac{rb}{\bar{h}}\frac{\partial s}{\partial x} + \frac{D}{a}\frac{\partial s}{\partial x} - D\frac{\partial^2 s}{\partial x^2} + s\frac{r}{h_0} = 0 \tag{5.25}$$

This is a linear differential equation of s with x-dependent coefficients. The relative importance of the terms in Equation 5.25, and in particular the rainfall (and evaporation) terms, is analyzed through scaling of the terms. The parameters are transformed into non-dimensional parameters in a way that the non-dimensional salinity and its derivatives are of the order unity, hence the coefficients determine the relative importance of the terms. The following transformations are made:

$$t' = t/T$$

$$x' = x/L$$

$$Q' = Q_r/Q_0$$

$$r' = r/r_0$$

$$D' = D/D_0$$

where Q_0 is the dry season fresh water discharge, r_0 is the net rainfall rate during the dry season (negative in case of net evaporation), and L is the tidal average salt intrusion length that corresponds to Q_0. With these transformations, Equation 5.25 can be scaled to:

$$r_S \frac{\partial s'}{\partial t'} + N_1 r' s' + (N_2 r' + N_3 Q' + N_4 D')\frac{\partial s'}{\partial x'} + N_5\left(\frac{\partial D'}{\partial x'}\frac{\partial s'}{\partial x'} + D'\frac{\partial^2 s'}{\partial x'^2}\right) = 0 \tag{5.26}$$

where the following non-dimensional coefficients determine the relative importance of the terms with respect to the rate of change of the salinity, $\partial s'/\partial t'$.

$$N_1 = r_0 T/h_0$$

$$N_2 = -r_0 Tb/(h_0 L)$$

$$N_3 = Q_0 T/(A_0 L \exp(-x' L/a))$$

$$N_4 = D_0 T/(aL)$$

$$N_5 = -D_0 T/L^2$$

where N_1 and N_2 weigh the impact of the net rainfall r on the rate of change of the salinity, N_1 through dilution and N_2 through advection; N_3 weighs the impact of the river discharge on the rate of change of the salinity; and N_4 and N_5 weigh the importance of the dispersion at the downstream boundary to the rate of change of the salinity. In Table 5.1, the values of the coefficients are presented for the estuaries studied.

Table 5.1 Non-dimensional coefficients determining the relative importance of terms to salt intrusion

Estuaries	h (m)	a (km)	b	A_0' (m^2)	L (km)	Q_0 (m^3/s)	r_0T (mm)	D_0	N_1 10^{-3}	N_2 10^{-3}	N_3[1] 10^{-3}	N_4 10^{-3}	N_5 10^{-3}
Mae Klong	5.2	102	155	1400	26	−30	−1.9	190	−0.37	2.18	−41.6	3.18	−12.48
Solo	9.2	226	226	2070	35	−10	−1.9	238	−0.21	1.33	−6.6	1.34	−8.63
Lalang	10.6	217	96	2550	65	−50	−2.0	892	−0.19	0.28	−15.6	2.81	−9.37
Limpopo	7	100	50	1340	60	−5	−2.0	145	−0.29	0.24	−3.7	1.07	−1.79
Tha Chin	5.3	87	87	1380	70	−10	−1.9	273	−0.36	0.45	−6.9	1.99	−2.47
Chao Phya	7.2	109	109	4300	50	−30	−1.9	332	−0.26	0.58	−7.8	2.70	−5.90
Incomati	2.9	42	42	1750	50	−1	−1.8	9	−0.62	0.52	−0.9	0.19	−0.16
Pungué	3.5	21	21	28,000	70	−20	−2.4	138	−0.69	0.21	−2.4	5.17	−1.25
Maputo	3.6	16	16	6460	40	−10	−1.8	105	−0.50	0.20	−6.0	7.28	−2.91
Thames	7.1	23	23	58,500	90	−20	−1.0	77	−0.14	0.04	−1.2	1.65	−0.42
Corantijn	6.5	64	64	34,600	50	−500	−2.0	230	−0.31	0.39	−19.0	3.19	−5.08
Sinnamary	3.8	39	39	1210	16	−100	−2.4	560	−0.63	1.54	−281.6	39.85	−97.13
Gambia	8.7	121	121	27,200	300	−2	−2.4	200	−0.28	0.11	−0.04	0.24	−0.10
Schelde	10.5	28	28	150,000	110	−90	−1.0	264	−0.10	0.02	−1.7	3.81	−0.97
Delaware	6.6	41	41	255,000	140	−300	−1.0	312	−0.15	0.04	−2.1	2.41	−0.71

[1] $x' = 0.5$

In evaluating the importance of the terms of Equation 5.26, the following should be taken into account. Firstly, although s' varies between 0 and 1, this is not the case for its derivatives. Secondly, the coefficient N_3 depends on x', and, hence, is not a constant. The choice was made to incorporate the factor $\exp(-x'L/a)$ in N_3 (instead of leaving it in the equation) to make the variables independent of the estuary under consideration, and to let the estuary-dependent coefficients determine the relative importance of the terms.

Consequently, a certain value of x' should be selected to allow comparison between estuaries. At the downstream boundary, where $x' = 0$, the salinity is determined by the ocean salinity, and evaporation has no effect. At the upstream boundary, where $x' = 1$, the salinity equals the fresh water salinity and the effect of evaporation is negligible. The part of the estuary where evaporation has the largest effect is in the central part of the salinity zone, where $x' = 0.5$, $D' \approx 0.5$, $S' \approx 0.5$, $\partial D'/\partial x' \approx -1$ and $\partial s'/\partial x' \approx -1$. Hence $x' = 0.5$ has been selected for the comparison.

To evaluate the importance of rainfall and evaporation in relation to other terms, the non-dimensional coefficients are grouped (see Table 5.2). In combining coefficients, it is assumed that $D' \approx 0.5$, $S' \approx 0.5$, $\partial D'/\partial x' \approx -1$ and $\partial s'/\partial x' \approx -1$, and that the magnitude of $D'\partial^2 s'/\partial x'^2$ is small when compared to the others. These assumptions are not exact since the values depend on the type of the salt intrusion curve: in case of a dome-shaped intrusion curve, $s' > 0.5$, $\partial s'/\partial x' > -1$,

Table 5.2 Non-dimensional coefficients indicating the relative importance of rainfall and evaporation in relation to discharge and dispersion

Estuaries	N_r 10^{-3}	N_Q 10^{-3}	N_D 10^{-3}	N_r/N_Q 10^{-3}	N_r/N_D
Mae Klong	−2.4	41.6	−15.1	−0.1	0.2
Solo	−1.4	6.6	−9.3	−0.2	0.2
Lalang	−0.4	15.6	−10.8	−0.0	0.0
Limpopo	−0.4	3.7	−2.3	−0.1	0.2
Tha Chin	−0.6	6.9	−3.5	−0.1	0.2
Chao Phya	−0.7	7.8	−7.2	−0.1	0.1
Incomati	−0.8	0.9	−0.3	−0.9	3.3
Pungué	−0.5	2.4	−3.3	−0.2	0.2
Maputo	−0.5	6.0	−6.6	−0.1	0.1
Thames	−0.1	1.2	−1.2	−0.1	0.1
Corantijn	−0.5	19.0	−5.7	−0.0	0.1
Sinnamary	−1.9	281.6	−117.0	−0.0	0.0
Gambia	−0.2	0.0	−0.2	−6.6	1.1
Schelde	−0.1	1.7	−2.9	−0.0	0.0
Delaware	−0.1	2.1	−1.9	−0.1	0.1

and $\partial^2 s'/\partial x'^2 < 0$; in case of a recession-shaped intrusion curve $s' < 0.5$, $\partial s'/\partial x' < -1$, and $\partial^2 s'/\partial x'^2 > 0$. They are, however, sufficiently accurate to evaluate the relative importance of the terms in orders of magnitude. This results in three coefficients: one coefficient N_r, which weighs the overall importance of rainfall:

$$N_r = N_1/2 - N_2$$

another coefficient which weighs the importance of fresh water inflow:

$$N_Q = -N_3$$

and a third coefficient which weighs the overall impact of the dispersion:

$$N_D = -0.5N_4 + N_5$$

In Table 5.2 these coefficients are presented, as well as the ratios of N_r to N_Q and N_D respectively. It can be concluded from Table 5.2 that the effect of evaporation is important in the Gambia, that there is some influence during minimum flow in the Incomati, but that the influence is not significant in the other estuaries studied.

5.4 TIME SCALES AND CONDITIONS FOR STEADY STATE

The assumption made in Equation 5.16 to arrive at the steady state equation for conservation of mass, requires that in the estuary, an equilibrium condition is reached between, on the one hand, advective salt transport through the downstream flushing of salt by the fresh water discharge, and, on the other hand, the full range of mixing processes induced by the tidal movement and the gravitational circulation. The time required for equilibrium to occur depends on 1) the rate at which the boundary conditions vary, in particular the rate of change of the fresh water discharge, and 2) the time required for the estuary system to adjust itself to a new situation. During the dry season, when the problem of salt intrusion is most acute, the variation of the fresh water discharge is generally slow. The question is: does the system react at the same pace as the boundary conditions, or does it need more time? In the latter case, the system lags behind steady state.

The estuary system reacts quite differently to an increase and to a decrease in the fresh water discharge. Generally the estuary reacts relatively quickly to an increase of the discharge. The new volume added at the upstream end of the estuary propagates as a mass wave through the system. However, the reaction to a decrease in the fresh water discharge is slow, since the process of salinization, gradually replacing the fresh water by saline water through mixing, takes time. In this respect, Van Dam and Schönfeld (1967) remark about the Schelde and Eems estuaries in The Netherlands that 'the characteristic time for reaching a new equilibrium, for example from a wet period to a dry period, is in the order of one year for the

Schelde, so that a final state is usually never reached.' Although for the Schelde this observation will prove to be somewhat exaggerated, in the Gambia, the process of adjustment to a reduction of the fresh water discharge is so slow that the salt intrusion always lags far behind steady state. This phenomenon is important. In the Gambia, it is due to the slowness of the mixing process that so much fresh water remains available in the upper part of the estuary, during the dry season. If equilibrium were reached, the salt intrusion would go much further upstream.

It is important to investigate in a given estuary how quickly the system adjusts to a new situation. If the time required for the system to reach equilibrium is too long in relation to the variation of the boundary conditions, then a steady state model may not be used.

System response time scale

Kranenburg (1986) developed a time scale for system response to a variation in discharge, which is based on the comparison of a steady state model with an unsteady state model. If the salinity difference per unit of time between subsequent steady states at a certain point along the estuary axis (as a result of a change in fresh water discharge) is defined as $\Delta S_{\mathrm{TA}}/\Delta t$, then a steady state is not reached as long as the mean tidal salinity adjustment rate $\partial s_{\mathrm{TA}}/\partial t$ is smaller, in absolute terms, than $\Delta S_{\mathrm{TA}}/\Delta t$, for an infinitely small time step Δt. Hence a condition for steady state is that:

$$\left|\frac{\partial s_{\mathrm{TA}}}{\partial t}\right| \geq \lim_{\Delta t \to 0}\left|\frac{\Delta S_{\mathrm{TA}}}{\Delta t}\right| \tag{5.27}$$

However, when the unsteady salinity variation is larger than the variation between subsequent steady states, the unrealistic situation occurs where the system reacts stronger than the forces driving the system. Hence, for the state of equilibrium to be reached, it is sufficient to assume that the unsteady state salinity variation should be approximately equal to the variation between steady states:

$$\frac{\partial s_{\mathrm{TA}}}{\partial t} \approx \lim_{\Delta t \to 0} \frac{\Delta S_{\mathrm{TA}}}{\Delta t} \tag{5.28}$$

In the following elaboration of Equation 5.28, for reasons of simplicity, the subscript TA for the tidal average situation in S, s, A, and D is removed. Combination with Equation 5.15 yields:

$$\frac{\partial}{\partial x}\left(Q_{\mathrm{f}} s - AD\frac{\partial s}{\partial x}\right) \approx -A \lim_{\Delta t \to 0} \frac{\Delta S}{\Delta t} \tag{5.29}$$

In fact, the integral of Equation 5.29 with respect to x represents the balance between, on the one hand, the salt flux, see Equation 5.4, which the system can produce, and on the other hand, the amount of salt per unit of time required to follow up subsequent steady states.

Since the time dependency of the steady state salinity S is merely through Q_f:

$$\lim_{\Delta t \to 0} \frac{\Delta S}{\Delta t} = \frac{\partial S}{\partial Q_f} \frac{dQ_f}{dt} \tag{5.30}$$

The most interesting case is the situation of extreme salt intrusion during the dry season, when the discharge is gradually diminishing. During the dry season, the depletion of the fresh discharge is exponential, following the normal recession curve used in hydrology:

$$\frac{dQ_f}{dt} = -\frac{Q_f}{T_Q} \tag{5.31}$$

where T_Q is the time scale of the discharge reduction (which corresponds to the residence time of the renewable groundwater in the river basin).

Integration of Equation 5.29 with respect to x, under the boundary condition that $s = 0$ at $x = L$, and combination with Equations 5.30, 5.31, and 5.16, disregarding S_f, yields:

$$(S - s) - \frac{A}{Q_f}\left(D_{SS}\frac{\partial S}{\partial x} - D\frac{\partial s}{\partial x}\right) \approx \frac{1}{T_Q}\int_x^L A\frac{\partial S}{\partial Q_f}dx \tag{5.32}$$

where D_{SS} refers to the steady-state dispersion of Equation 5.16 and D to the unsteady-state dispersion. Depending on the theory used, the tidal average dispersion coefficient D is a function of x, Q_f, $\partial S/\partial x$, S or any combination of them. If the dispersion coefficient was only a function of x and Q_f, then the two coefficients would be equal. If, however, the dispersion coefficient is assumed to depend on S or $\partial S/\partial x$, as is the case in several theories, then they can be different, although percentage-wise they cannot differ much. However, since also the salinity gradients between steady state and unsteady state are different, the dispersion terms are different, and do not annihilate.

Kranenburg solves this problem in the following way. He observes that $(s - S) = 0$ both at $x = 0$ and at $x = L$, and that somewhere in the middle of the estuary a point should lie where $(s - S)$ reaches a maximum value, and where, hence, $\partial s/\partial x = \partial S/\partial x$. With $D \approx D_{SS}$, this implies that there is a point $x = X$ where the steady state and unsteady state dispersion terms of Equation 5.23 annihilate each other whether or not D depends on $\partial S/\partial x$. Hence:

$$(S - s)\Big|_X \approx \frac{1}{T_Q}\int_X^L A\frac{\partial S}{\partial Q_f}dx \tag{5.33}$$

Subsequently, the time scale for system response is defined as:

$$T_K = \frac{1}{S(X)} \int_X^L A \frac{\partial S}{\partial Q_f} dx \tag{5.34}$$

Combination with Equation 5.33 yields:

$$T_K \approx T_Q \frac{(S-s)}{S}\bigg|_X \tag{5.35}$$

Assuming that the maximum difference between steady and unsteady state salinity lies at about half the salt intrusion length ($X=L/2$). This implies that to reach a state of 90 percent equilibrium at $x=L/2$, $(S-s)/S=10$ percent, and T_K should not be more than 0.1 T_Q. To reach a further state of equilibrium, T_K should be less.

A more simple approach to determine the system response time is on the basis of the time required for the system to adjust itself to a new steady state S. The time scale is defined as:

$$\frac{\partial s}{\partial t} = \frac{S-s}{T_S} \tag{5.36}$$

Analogous to the operations carried out to obtain T_K, Equation 5.36 is modified into:

$$Q_f(S-s) - A\left(D_{SS}\frac{\partial s}{\partial x} - D\frac{\partial S}{\partial x}\right) \approx -\frac{1}{T_S}\int_x^L A(S-s)dx \tag{5.37}$$

If Kranenburg's assumption is applied that at $x=X$: $D\partial s/\partial x = D_{SS}\partial S/\partial x$, then Equation 5.37 yields:

$$Q_f T_S(S-s) \approx -\int_X^L A(S-s)dx \tag{5.38}$$

Because the spatial distribution of s is not known, it is assumed that in reaching the new steady state, the difference in salinity $(S-s)$ is proportional to S. In the reach between X and L, this seems a reasonable assumption:

$$T_S \approx -\frac{1}{Q_f S(X)}\int_X^L AS dx \tag{5.39}$$

This is the time required for the fresh water discharge, with a salinity $S(X)$, to replace the salt accumulated upstream of X. For a steady state to occur, this time scale should not be larger than the time scale of the discharge reduction.

Particle travel time

A good time scale to compare T_S or T_K with is the average time T_f required for a fresh water particle to travel over the salt intrusion length:

$$T_f = -\int_0^L \frac{A}{Q_f} dx \tag{5.40}$$

One could call this the flushing time scale, since it equals the time required to flush the estuary with fresh water. In an exponentially shaped channel with a convergence length a, Equation 5.40 yields:

$$T_f = -\frac{A_0 a}{Q_f}\left(1 - \exp\left(-\frac{L}{a}\right)\right) \tag{5.41}$$

If the exponential function describing the cross-sectional area consists of two branches (as we saw in Chapter 2, it sometimes happens), the upstream branch is considered most important since for the determination of T_S integration is done between X and L, and consequently in Equation 5.41 the convergence length of the upstream branch is used and a value A_0' instead of A_0 which is obtained by extrapolation of the upstream branch to the estuary mouth. This value has been used for the computation of T_f in Table 5.6.

The ratio of T_S to T_f is an interesting parameter for further study, since it no longer depends on the fresh water discharge Q_f, and hence only includes geometric variables of the estuary under study.

Given a mathematical model for the steady state salinity distribution, Equation 5.39 can be solved, and the system time scale T_S determined. This is done in Section 5.6, after we have obtained an expression for the salt intrusion length L.

5.5 PREDICTIVE MODEL FOR STEADY STATE

5.5.1 Expressions for HWS, LWS, and TA

Steady state salt intrusion models can be divided into three types, depending on their derivation: low water slack (LWS) models, e.g. Ippen and Harleman (1961); tidal average (TA) models, e.g. Van den Burgh (1972); and high water slack (HWS) models, e.g. Savenije (1989). A LWS model is calibrated on measurements carried out at LWS and a HWS model with measurements carried out at HWS. For the calibration of a TA model, the average salinity has to be derived from measurements carried out during a full tidal cycle at several points along the estuary axis, which, if done well, is very elaborate and time consuming.

Although HWS models are not commonly used, they are the most practical models because the best time instant to carry out a salt intrusion measurement is at HWS; this is for the following reasons:

1. The moment that HWS occurs is easily determined. The observer measures the salinity, when the in-going current slacks. Although the same advantage

applies to LWS, the accessibility at LWS is generally poor. The estuary is shallow and sometimes, it is very difficult to reach the waterside. Inaccessible mud flats often separate the river from the banks.

2. If the salinity at the downstream boundary is not known, it can easiest be estimated at HWS. At HWS, the salinity at the estuary mouth is generally equal or almost equal to the sea salinity, which is not, or not much, affected by the fresh water discharge from the estuary.
3. At HWS, the salt intrusion is at its maximum. Generally, it is the maximum intrusion that is of interest to planners.
4. A single observer in a small outboard driven boat can travel with the tidal wave and measure the entire salt intrusion curve at HWS. If the intrusion length is not too long, he may even return to the estuary mouth and repeat the measurement for LWS.

As it seems logical to assume that the TA salt intrusion is directly related to the fresh water discharge, many predictive models are of the TA type. These models, however, do not provide information on the maximum and minimum salinity reached at a certain location as a function of the tide. This is especially important, when during part of the day, the water is fresh, while during another part of the day, the water is brackish. In this section, it will be shown that the model can be calibrated on HWS—which has all the advantages mentioned above—can be used to compute the situation at LWS and TA, and that the calibration parameters obtained can be well related to hydrological, hydraulic, and geometrical parameters.

In Section 5.2 it has been shown that the same type of Equation 5.17 can be used to describe the three situations of HWS, LWS, and TA. Because in a steady state situation, $\partial S/\partial t = 0$, and hence $\partial S/\partial x = \mathrm{d}S/\mathrm{d}x$, no partial derivatives are used in the subsequent derivations in this section. Equation 5.17 can be combined with Van den Burgh's equation, described in Chapter 4, and the geometric assumptions, described in Chapter 2, to yield analytical equations for HWS, LWS, and TA that are mathematically related. In the following paragraphs, relations are derived between the salinity distributions at HWS, LWS, and TA. These relations allow the determination of the dispersion at LWS and TA, and the corresponding salinity distributions on the basis of the calibrated dispersion at HWS.

To this end, in agreement with Equations 5.9, 5.10, and 5.16, Equation 5.17 is expanded to read:

$$S_i - S_\mathrm{f} = \frac{A}{Q_\mathrm{f}} D_i \frac{\mathrm{d}S_i}{\mathrm{d}x} \tag{5.42}$$

where A is the tidal average cross-sectional area, $D_i = D_i(x)$ is the dispersion coefficient at HWS, LWS, and TA respectively. It should be noted that the values of Q_f and A are assumed to be the same for HWS, LWS, and TA. The error made by using the same value for A is compensated by D_i, which then also incorporates the effect of a smaller cross-sectional area at LWS and a larger cross-sectional area at HWS. In this way, all the differences between HWS, LWS, and TA are incorporated in one variable D_i.

In Chapter 4, the Van den Burgh equation has been presented. Van den Burgh assumed that his relation, which had been derived for the TA situation, could be used for LWS and HWS as well: he assumed that the salt intrusion curve obtained for the TA situation could be shifted upstream over half the tidal excursion to obtain the HWS situation, and downstream over half the tidal excursion to obtain the LWS situation. This is in agreement with what was observed by O'Kane (1980) in his oscillating framework approach (see Section 5.2) and by Park and James (1990) who stated that the instantaneous—not tidal average—salt flux F was found to arise predominantly from the product of the tidal average salinity S and the tidal velocity U. This means that the tidal average salinity variation represented by the terms in Equation 5.15 is small when compared to the instantaneous advection $UA\partial S/\partial x$. Hence, over one tidal cycle, a short time compared to the time required to adjust the terms in Equation 5.15, it is justified to assume average mixing conditions corresponding to the TA situation.

Hence Van den Burgh's expression is used for HWS, LWS and TA:

$$\frac{\mathrm{d}D_i}{\mathrm{d}x} = K\frac{Q_\mathrm{f}}{A} \tag{5.43}$$

Combination of Equation 5.42 with Equation 5.43 yields:

$$\frac{\mathrm{d}S}{S - S_\mathrm{f}} = \frac{1}{K}\frac{\mathrm{d}D}{D} \tag{5.44}$$

where the subscript i has been disregarded for the sake of simplicity, but it is understood that Equation 5.44 can be applied to the HWS, LWS, and TA situation. Integration results in:

$$\frac{S - S_\mathrm{f}}{S_0 - S_\mathrm{f}} = \left(\frac{D}{D_0}\right)^{\frac{1}{K}} \tag{5.45}$$

where S_0 and D_0 are boundary conditions at $x=0$ for the HWS, TA or LWS condition. In addition, integration of Equation 5.43 in combination with an exponentially varying cross section yields:

$$\frac{D}{D_0} = 1-\beta\left(\exp\left(\frac{x}{a}\right) - 1\right) \tag{5.46}$$

where:

$$\beta = -\frac{KaQ_f}{D_0 A_0} \tag{5.47}$$

where β is the dispersion reduction rate, which is always positive (since Q_f is negative), determining the longitudinal variation of D. From Equation 5.45, it can be seen that $S = S_f$ when $D = 0$. Since $S = S_f$ at $x = L$, the intrusion length, Equation 5.46 can be elaborated to yield an expression for the intrusion length:

$$L = a\ln\left(\frac{1}{\beta} + 1\right) \tag{5.48}$$

Since β is positive, the argument of the natural logarithm is always greater than unity.

Application of Equations 5.45 and 5.46 to the HWS situation, disregarding S_f for reasons of convenience, yields:

$$\frac{S^{HWS}}{S_0^{HWS}} = \left(\frac{D^{HWS}}{D_0^{HWS}}\right)^{\frac{1}{K}} \tag{5.49}$$

$$\frac{D^{HWS}}{D_0^{HWS}} = 1 + \frac{KaQ_f}{D_0^{HWS} A_0}\left(\exp\left(\frac{x}{a}\right) - 1\right) \tag{5.50}$$

Since the tidal excursion may be assumed to be independent of x, the envelope lines of the salinity at HWS and LWS should have the same shape. The line at TA is obtained through a horizontal translation over a distance equal to half the tidal excursion. Hence:

$$S_0^{TA} = S^{HWS}(E/2) \tag{5.51}$$

$$S^{TA}(E/2) = S^{HWS}(E) \tag{5.52}$$

Application of the combined Equations 5.9 and 5.10 at $x = E/2$ and $x = E$, division of the results, some elaboration, and substitution of Equations 5.51 and 5.52 yields:

$$\frac{S^{HWS}(E)}{S^{HWS}(E/2)} = \frac{S^{TA}(E/2)}{S_0^{TA}} = \left(1 + \frac{KaQ_f}{A_0 D^{HWS}(E/2)} \frac{\exp\left(\frac{E}{2a}\right) - 1}{\exp\left(-\frac{E}{2a}\right)}\right)^{\frac{1}{K}} \tag{5.53}$$

If one uses Equations 5.45 and 5.46 for TA, and substitutes $x=E/2$, then comparison of the result with Equation 5.53 shows that:

$$D_0{}^{TA} = D^{HWS}(E/2)\exp\left(-\frac{E}{2a}\right) \tag{5.54}$$

Similarly for LWS, one obtains:

$$S_0^{LWS} = S^{HWS}(E) \tag{5.55}$$

$$D_0{}^{LWS} = D^{HWS}(E)\exp\left(-\frac{E}{a}\right) \tag{5.56}$$

Estuaries with a complex geometry

In some estuaries, the geometry cannot be described by a single exponential function. Several estuaries require two branches to describe the longitudinal variation of A, see for instance the Incomati estuary of Figure 2.8. In fact, there are many estuaries like that. In Table 5.3 there are: the Limpopo, the Tha Chin, the Incomati, the Pungué, the Maputo, the Corantijn, and the Sinnamary. In fact most of the estuaries have two branches, often accounting for a trumpet shape near the estuary mouth. In those estuaries, the above equations can be solved stepwise. The geometry is given by:

$$A = A_0 \exp\left(-\frac{x}{a_1}\right) \quad \text{if} \quad 0<x<x_1 \tag{5.57}$$

$$A = A_1 \exp\left(-\frac{x-x_1}{a_2}\right) \quad \text{if} \quad x>x_1 \tag{5.58}$$

where x_1 is the inflection point, $A_1=A(x_1)$, a_1 is the convergence length of the downstream branch, and a_2 of the upstream branch. In general, the downstream convergence length is shorter than the upstream one, resulting in the said trumpet shape.

The Equations 5.45 and 5.46 can be applied normally for the downstream part, using a_1 for a. Subsequently, a value for the dispersion at the inflection point D_1 needs to be determined, as well as the salinity at the inflection point: S_1. This is done by substitution of x_1 for x in Equations 5.45–5.47, using a_1 for a. Subsequently, Equations 5.45 and 5.46 can be used for the upstream end, taking D_1 and S_1 as the downstream boundary conditions. The intrusion length can be computed as follows:

$$L^{HWS} = x_1 + a_2 \ln\left(\frac{1}{\beta_1}+1\right) \tag{5.59}$$

where β_1 is defined by:

$$\beta_1 = -\frac{Ka_2Q_f}{D_1A_1} \tag{5.60}$$

using the value of D_1 determined at the inflection point.

After calibration of the model on HWS observations, the values of S_0^{HWS} and D_0^{HWS} are known. Consequently, the salinities and dispersion coefficients at any point along the estuary for HWS can be calculated using Equations 5.49 and 5.50.

By substitution of these HWS values into Equations 5.51, 5.54–5.56, the TA salinities and the LWS salinities can be computed using Equations 5.45 and 5.46 for TA and LWS, respectively.

In the above equations, there remain two unknown model parameters: K and D_0^{HWS}. In addition, there is another variable that is often unknown: the fresh river discharge Q_f, which in an estuary is one of the most difficult parameters to determine. Fortunately, in the above equations, Q_f always occurs in the same term as the dispersion coefficient, which permits them to be combined into only one variable, the mixing coefficient α (m^{-1}):

$$\alpha = -\frac{D}{Q_f} \tag{5.61}$$

The value of α at the estuary mouth, α_0, is a model parameter that can be obtained through calibration. The two remaining calibration parameters function differently: K is a value that is fixed for a certain estuary, whereas α_0 varies over time, responding to the tidal range and the river discharge. They affect the equations differently and can be easily found by fitting the model to observations.

To be able to determine the fresh water discharge from α_0, an additional relation is needed between α_0 and Q_f. This relation, which is required to make the model predictive, is established in the next section.

Some applications as illustration

The steady state model has been applied in numerous estuaries for a wide range of river flows and tidal ranges. The data of these measurements are summarized in Table 5.3 In Figure 5.2, the example of the Maputo estuary in Mozambique was presented already. Here, some more illustrations are given of the steady state model applied to three other Mozambican estuaries: the Pungué (Figure 5.3), the Incomati (Figure 5.4) and the Limpopo (Figure 5.5). These measurements were conducted by different persons: the ones in the 1980s by the author, the ones in the 1990s by H.A. Zanting, and those in 2002 by S. Graas.

It can be seen that the model performs very well, although the estuaries are very different in character. The Pungué has a 'dome-shaped' intrusion curve, the Limpopo has a 'recession-shaped' intrusion curve, and the Incomati and the

Maputo have a 'bell-shaped' intrusion curve. The toes of the curves are also different. The toe of the Incomati is very flat, corresponding with a very small value of K ($K=0.15$), whereas the Limpopo and Pungué have a steeper toe ($K=0.5$ and 0.3 respectively).

When calibrating the model to measurements, one uses different values of K and α_0. Because these two parameters affect the fit in different ways, it is generally possible to find a satisfactory combination of K and α_0. One should realize that, K should be independent of the river discharge and the tide, whereas α_0 should vary with river discharge and tide. So, if observations during different flow regimes are collected, then different values of α_0 should be obtained using the same value of K. This can also be seen in Table 5.3.

In general, it is important that sufficient measurements along the estuary axis are taken, because individual measurements can have considerable errors due to: timing errors, variation of salinity over the cross section, and local mixing effects (trapping etc.). The best way to derive a longitudinal distribution is to travel by boat during HWS and LWS. But this may require a very fast boat in some estuaries, since the tidal wave can travel fast (see Section 3.2).

5.5.2 Empirical relations for the predictive model

To turn the steady state model into a predictive model, (semi-) empirical relations are required that relate the two calibration parameters K and α_0 to hydrodynamic and geometrical bulk parameters. These bulk parameters are dimensionless numbers composed of geometrical (a, b, A_0, B_0, h_0), hydrological (Q_f) and hydraulic (H, E, C, υ) parameters that influence the process of mixing and advection. In the past, several researchers sought significant bulk parameters to be used for predicting model parameters from directly measurable physical quantities. In the following, there is a short review of the empirical work by Rigter (1973), Fischer (1974), Van den Burgh (1972), and Van Os and Abraham (1990). With the exception of Van den Burgh, all these investigators based their analysis on laboratory tests and prototype measurements in estuaries with constant cross section.

Classical approaches

Combination of Equations 5.47 and 5.48 yields an explicit analytical relation between the salt intrusion length L and the effective dispersion at the estuary mouth D_0, or, for that matter, between L and α_0 if Equation 5.61 is used. Prominent in this equation is the presence of the convergence length and the logarithmic function; both stemming from the exponential shape of alluvial estuaries. Classical literature on the intrusion length, however, is almost entirely based on prismatic channels. This may be difficult to understand in hindsight, since natural, alluvial, estuaries are never prismatic, but it is not so strange if we consider the reasons behind it. Pioneers in the development of formulas to predict the intrusion length were the U.S. Army Corps of Engineers, and the Dutch Ministry of Public Works, both of which were involved in the design of shipping access

Table 5.3 Measured salinity distributions and calibrated values of K and α_0 for different estuaries

Estuary	Date	T (s)	H_0 (m)	E_0 (km)	Q_f (m^3/s)	S_0 (kg/m^3)	f	K	a_1 (km)	x_1 (km)	a_2 (km)	A_0 ($10^3 m^2$)	h (m)	N	F^2	F_d	a (km)	α_0 (m^{-1})	α_0' (m^{-1})	D_0 (m^2/s)	D_0' (m^2/s)
Mae	20/01/77	86,400	2	11	120	28	0.028	0.3			102	1.4	5.20	0.673	0.003	0.157		7.2	5.4	864	647
Klong	08/03/77	44,400	1.5	10	60	30	0.028	0.3			102	1.4	5.20	0.190	0.010	0.458		6.2	5.4	372	324
	09/04/77	44,400	2	14	36	29	0.028	0.3			102	1.4	5.20	0.082	0.019	0.929		9.0	8.1	324	292
Solo	26/07/88	86,400	0.8	9	50	35	0.023	0.6			226	2.07	9.20	0.232	0.001	0.047		9.2	7.4	460	369
	08/09/88	86,400	0.4	5	7	35	0.023	0.6			226	2.07	9.20	0.058	0.000	0.015		12.5	15.7	88	103
Lalang	20/10/89	86,400	2.6	27	120	25	0.023	0.7			217	2.55	10.60	0.151	0.009	0.519		9.0	8.1	1080	970
Limpopo	04/04/80	44,440	1.1	7	150	30	0.026	0.5	50.4	20	130	1.71	7.00	0.557	0.004	0.166	58	9.0	7.7	1350	1158
	31/12/82	44,440	1.1	8	2	35	0.026	0.5	50.4	20	130	1.71	7.00	0.006	0.005	0.186	105	38.0	42.3	76	85
	22/04/83	44,440	0.5	4	1	33	0.026	0.5	50.4	20	130	1.71	7.00	0.006	0.001	0.049	108	45.0	40.2	45	40
	24/07/94	44,440	0.9	6.8	5	35	0.026	0.5	50.4	20	130	1.71	7.00	0.019	0.003	0.135	101	25.0	25.7	125	128
	10/08/94	44,440	1.0	7.1	3	35	0.026	0.5	50.4	20	130	1.71	7.00	0.011	0.004	0.147	103	29.0	33.1	87	99
Tha	16/04/81	86,400	1.6	12	55	26	0.039	0.4	22	22	87	3	5.30	0.132	0.004	0.197	55	15.5	10.4	853	572
Chin	27/02/86	44,400	2.6	20	40	31	0.039	0.4	22	22	87	3	5.30	0.030	0.039	1.739	54	15.0	12.7	600	507
	01/03/86	86,400	1.8	14	40	34	0.039	0.4	22	22	87	3	5.30	0.082	0.005	0.205	61	16.5	13.7	660	550
	13/08/87	44,400	2	15	39	27	0.039	0.4	22	22	87	3	5.30	0.038	0.022	1.123	51	12.0	10.9	468	424
Chao	05/06/62	86,400	2.2	22	63	29	0.031	0.8			109	5.3	7.20	0.058	0.009	0.437		11.4	9.3	718	587
Phya	17/03/80	86,400	1.5	18	43	32	0.031	0.8			109	5.3	7.20	0.048	0.006	0.265		15.2	10.7	654	461
	28/03/80	86,400	1.7	20	31	34	0.031	0.8			109	5.3	7.20	0.031	0.007	0.308		17.7	13.7	549	425
	29/01/83	86,400	2.4	26	90	33	0.031	0.8			109	5.3	7.20	0.070	0.013	0.537		12.0	9.0	1080	813
	23/02/83	86,400	1.6	19	100	28	0.031	0.8			109	5.3	7.20	0.106	0.007	0.338		9.4	6.7	940	675
	16/01/87	86,400	2.5	15	180	20	0.031	0.8			109	5.3	7.20	0.241	0.004	0.295		5.4	3.8	972	680
Incomati	30/07/80	44,440	1.4	7	3	32	0.022	0.2	7.5	14	42	8.1	2.90	0.002	0.009	0.377	33	9.5	11.2	29	34
	05/09/82	44,440	1.4	7	2	35	0.022	0.2	7.5	14	42	8.1	2.90	0.002	0.009	0.344	35	16.0	13.5	32	27
	10/02/83	44,440	1.2	8	1	35	0.022	0.2	7.5	14	42	8.1	2.90	0.001	0.011	0.450	37	30.0	19.5	30	19
	23/06/93	44,440	1.4	8	4	35	0.022	0.2	7.5	14	42	8.1	2.90	0.003	0.011	0.450	32	10.0	11.0	40	44
	07/07/93	44,440	2.6	9	4	35	0.022	0.2	7.5	14	42	8.1	2.90	0.002	0.014	0.569	32	9.4	11.7	38	47

Pungué	26/09/80	44,440	6.3	20	22	34	0.031	0.3	21	38	12	26.4	3.50	0.002	0.058	2.397	16	15.0	10.8	308	238
	26/05/82	44,440	5	10	50	32	0.031	0.3	21	38	12	26.4	3.50	0.008	0.015	0.637	18	5.2	5.5	210	225
	06/08/82	44,440	5.2	14	36	34	0.031	0.3	21	38	12	26.4	3.50	0.004	0.029	1.175	17	5.3	6.6	191	237
	22/09/82	44,440	5.2	14	26	35	0.031	0.3	21	38	12	26.4	3.50	0.003	0.029	1.141	17	8.0	8.1	208	212
	29/10/82	44,440	6	16	60	34	0.031	0.3	21	38	12	26.4	3.50	0.006	0.037	1.534	17	5.0	5.4	300	326
	03/10/93	44,440	5.3	15	10	35	0.031	0.3	21	38	12	26.4	3.50	0.001	0.033	1.310	16	28.0	15.0	280	140
	12/10/93	44,440	3.8	15	10	35.5	0.031	0.3	21	38	12	26.4	3.50	0.001	0.033	1.292	17	15.0	13.5	150	135
	16/10/93	44,440	6.4	16	10	35.5	0.031	0.3	21	38	12	26.4	3.50	0.001	0.037	1.469	16	26.0	15.0	260	150
	31/01/02	44,440	6.2	20	262	28	0.031	0.3	21	38	12	26.4	3.50	0.022	0.058	2.911	19	1.9	2.4	498	622
	27/02/02	44,440	6.1	20	200	29	0.031	0.3	21	38	12	26.4	3.50	0.017	0.058	2.811	19	2.3	2.8	460	562
	01/03/02	44,440	6.7	24	150	28	0.031	0.3	21	38	12	26.4	3.50	0.011	0.084	5.192	18	2.8	3.7	420	549
Maputo	28/04/82	44,440	2.8	13	25	35	0.022	0.4	4	8	16	40	3.60	0.002	0.024	0.957	13	5.5	8.8	113	221
	15/07/82	44,440	1.5	6	8	35	0.022	0.4	4	8	16	40	3.60	0.001	0.005	0.204	14	7.2	10.0	58	80
	06/04/82	44,440	2.9	10	180	35	0.022	0.4	4	8	16	40	3.60	0.020	0.014	0.566	12	3.0	3.0	540	547
	19/04/82	44,440	3.3	12	120	35	0.022	0.4	4	8	16	40	3.60	0.011	0.020	0.815	12	3.6	5.0	432	478
	02/05/84	44,440	3.4	13	70	28	0.022	0.4	4	8	16	40	3.60	0.006	0.024	1.196	13	3.9	5.8	273	337
	17/05/84	44,440	3.3	14	50	31	0.022	0.4	4	8	16	40	3.60	0.004	0.028	1.253	13	5.0	6.2	200	311
	29/05/84	44,440	2.8	12	40	30	0.022	0.4	4	8	16	40	3.60	0.004	0.020	0.951	13	5.8	6.1	192	245
	02/08/84	44,440	2.8	11	49	31	0.022	0.4	4	8	16	40	3.60	0.005	0.017	0.773	13	3.9	5.6	191	273
Thames	07/04/49	44,440	5.3	14	40	33	0.026	0.2			23	58.5	7.10	0.002	0.014	0.597		6.3	9.0	252	359
Corantijn	02/07/65	44,440	2	11	1995	35	0.026	0.2	19	18	64	69	6.50	0.117	0.009	0.379	48	0.4	0.5	798	898
	16/08/65	44,440	2.2	12	680	35	0.026	0.2	19	18	64	69	6.50	0.036	0.011	0.451	53	0.8	0.7	544	498
	08/12/78	44,440	1.8	10	115	20	0.026	0.2	19	18	64	69	6.50	0.007	0.008	0.549	54	1.0	1.2	115	137
	14/12/78	44,440	2.3	12	130	19	0.026	0.2	19	18	64	69	6.50	0.007	0.011	0.832	55	1.2	1.2	156	153
	20/12/78	44,440	1.6	9	220	18	0.026	0.2	19	18	64	69	6.50	0.016	0.006	0.494	55	1.1	0.8	242	169
Sinnamary	12/11/93	44,440	2.6	10	148	26.5	0.031	0.5	2.64	3	39	3.5	3.80	0.188	0.013	0.708	28	3.4	5.6	503	685
	27/04/94	44,440	2.9	12	108	22.8	0.031	0.5	2.64	3	39	3.5	3.80	0.114	0.019	1.185	28	5.5	5.5	486	595
	02/11/94	44,440	2.7	11	106	29.1	0.031	0.5	2.64	3	39	3.5	3.80	0.122	0.016	0.780	32	6.5	5.2	689	556
	03/11/94	44,440	2.9	12	106	26.4	0.031	0.5	2.64	3	39	3.5	3.80	0.112	0.019	1.024	31	5.5	5.3	583	566
Delaware	23/08/32	44,440	1.7	8	120	32	0.026	0.2			41	255	6.60	0.003	0.005	0.216		1.2	0.9	144	111
	04/10/32	44,440	1.7	8	72	32	0.026	0.2			41	255	6.60	0.002	0.005	0.216		1.7	1.2	122	86

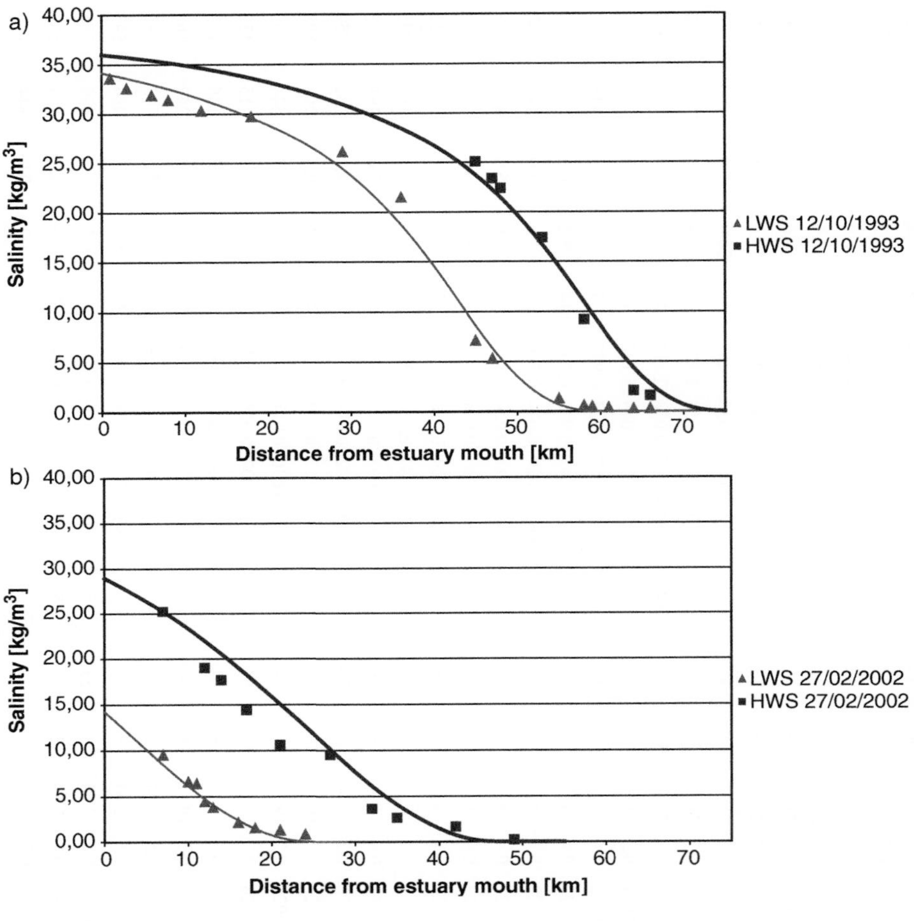

Figure 5.3 Measured and computed salt intrusion curves in the Pungué estuary: a) on 12/10/1993, b) on 27/2/2002. The draw lines are computed, the symbols indicate the measurements.

channels, waterways and harbors, particularly in the Mississippi delta and the Rotterdam Waterway. These channels, which were often man made and artificially kept at depth, had a prismatic, or near prismatic form. An important research question was what would happen to the salt intrusion if these channels were deepened. As a result, intensive laboratory experiments were conducted Waterways Experimental Station (WES) in Vicksburg, Mississippi, and at Delft Hydraulics in The Netherlands, all in flumes with a constant cross section.

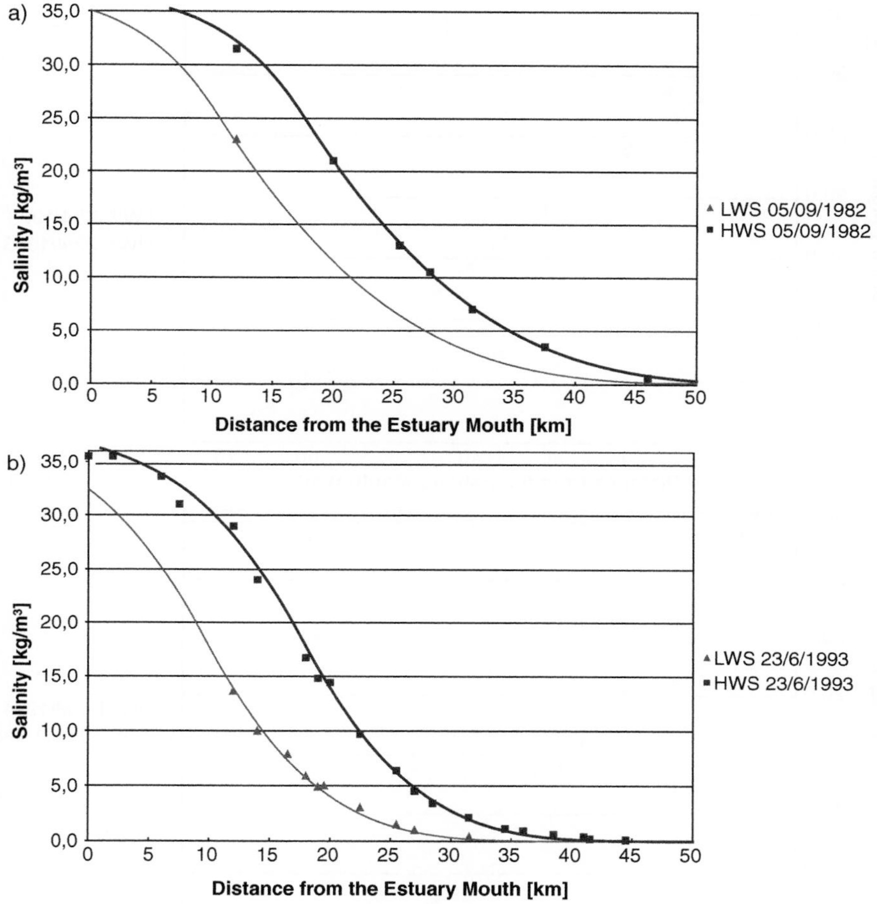

Figure 5.4 Measured and computed salt intrusion curves in the Incomati estuary: a) on 05/09/1982, b) on 23/06/1993. The draw lines are computed, the symbols indicate the measurements.

Since the early research on salt intrusion started in prismatic channels, this set the scene for further research, including analytical research which did not have the drawback of having to construct complicated flumes with a varying cross section. Apparently, it is difficult to change course once you are on a certain track. We shall see further on, however, that the formulae derived for prismatic channels perform very poorly for natural channels. But let us first see how the equations derived above perform under prismatic conditions.

In an estuary with constant cross section, D decreases linearly in upstream direction ($D = D_0 + KU_f\ x$), which follows directly from the integration of

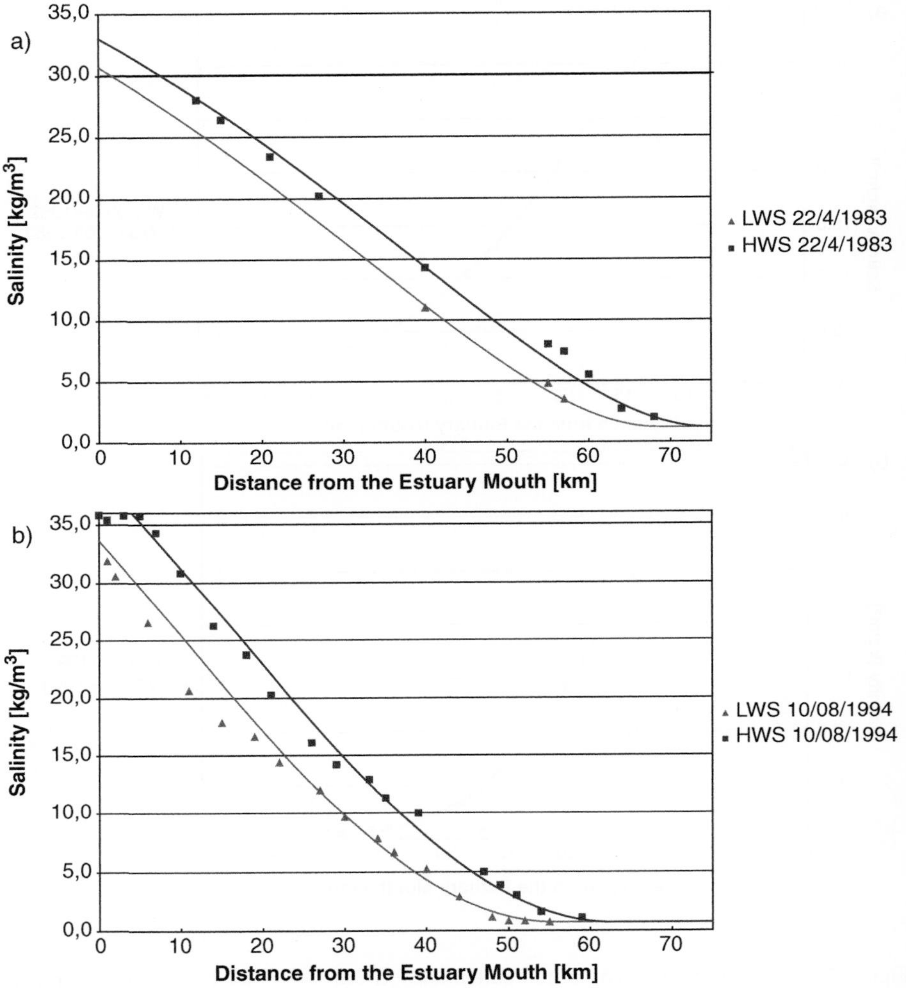

Figure 5.5 Measured and computed salt intrusion curves in the Limpopo estuary: a) on 22/04/1983, b) on 10/08/1994. The draw lines are computed, the symbols indicate the measurements.

Equation 5.43 with $A = A_0$. Substitution into Equation 5.45 and considering that $S = S_f$ for $x = L$ yields the following expression:

$$L = -\frac{D_0 A_0}{K Q_f} \tag{5.62}$$

If this equation is combined with an empirical relation for the intrusion length L, then an empirical relation is obtained for D_0 or α_0.

Rigter (1973), on the basis of flume data of the Waterways Experiment Station (WES), arrived at the following empirical relation:

$$L^{\mathrm{LWS}} = 1.5\pi \frac{h_0}{f_{\mathrm{D}}}\left(\boldsymbol{F}_{\mathrm{d}}^{-1} N^{-1} - 1.7\right) \approx 4.7 \frac{h_0}{f_{\mathrm{D}}} \boldsymbol{F}_{\mathrm{d}}^{-1} N^{-1} \tag{5.63}$$

where h_0 is the tidal average depth at the estuary mouth, f_{D} is the Darcy-Weisbach friction factor ($f_{\mathrm{D}} = 8g/C^2$), N is Canter-Cremers estuary number defined in Section 2.1 as the ratio of the fresh water entering the estuary during a tidal cycle ($-Q_{\mathrm{f}}T$) to the flood volume of salt water entering the estuary over a tidal cycle, P_{t}:

$$N = -\frac{Q_{\mathrm{f}}T}{P_{\mathrm{t}}} = -\frac{U_0 T}{E} = -\pi \frac{U_0}{\upsilon_0} \tag{5.64}$$

where $U_0 = U_{\mathrm{f}}(0) = Q_{\mathrm{f}}/A_0$ and υ_0 is the tidal velocity amplitude at the estuary mouth. In Equation 5.64, use has been made of Equations 2.65 and 2.74. Finally, the densimetric Froude number $\boldsymbol{F}_{\mathbf{d}}$ is defined as:

$$\boldsymbol{F}_{\mathrm{d}} = \frac{\rho \upsilon_0^2}{\Delta\rho g h_0} = \frac{\rho}{\Delta\rho} \boldsymbol{F}^2 \tag{5.65}$$

where $\boldsymbol{F}$ is the Froude number ($\boldsymbol{F} = \upsilon/\sqrt{(gh)}$). It is observed that, since in alluvial estuaries both N and $\boldsymbol{F}_{\mathbf{d}}$ are much smaller than unity, the number 1.7 in Equation 5.63 can be disregarded, and that Rigter's intrusion length is inversely proportional to N and $\boldsymbol{F}_{\mathbf{d}}$.

Fischer (1974), in a discussion of Rigter's results, and using the same data, derived the following formula:

$$L^{\mathrm{LWS}} = 17.7 \frac{h_0}{f_{\mathrm{D}}^{0.625}} \boldsymbol{F}_{\mathbf{d}}^{-0.75} N^{-0.25} \tag{5.66}$$

Van den Burgh (1972) made use of limited field observations in real estuaries: the Rotterdam Waterway, the Schelde, the Haringvliet (a tidal branch of the Rhine-Meuse delta) and the Eems. He reached at a quite similar relation as the earlier researchers. He found that the mean tidal dispersion at the mouth obeyed the following relation:

$$D_0{}^{\mathrm{TA}} = 26(Ng)^{0.5} h_0^{1.5} \tag{5.67}$$

In combination with Equations 5.62, 5.64, and 5.65 this yields for prismatic channels:

$$L^{TA} = -\frac{26h_0}{K}\frac{\sqrt{gh_0}}{\upsilon_0}\frac{\upsilon_0}{u_0}N^{0.5} = 26\pi\frac{h_0}{K}\boldsymbol{F}^{-1}N^{-0.5} \tag{5.68}$$

It is clear that, although Van den Burgh used $\boldsymbol{F}^2$ instead of $\boldsymbol{F_d}$, there is similarity between the methods presented; most importantly, they are all linear in h_0. Since the relative density difference $\Delta\rho$ over the intrusion length of a well-mixed estuary does not significantly vary from estuary to estuary—sea salinity being virtually the same everywhere ($\Delta\rho = 25\,\text{kg/m}^3$)—, and the same applies to the roughness, which has been said to be low and not much different in the various estuaries, the major difference between the methods lies in the exponents used for the two bulk parameters: the Froude number **F** and Canter-Cremers number N. This was also observed by Van Os and Abrahams (1990) who developed a formula similar to Rigter's for use at Delft Hydraulics:

$$L^{LWS} = 4.4\frac{h_0}{f_D}\boldsymbol{F_d}^{-1}N^{-1} \tag{5.69}$$

About the exponents in Equation 5.69, it can be observed that the exponent of N is negative since the salt intrusion length reduces if Q_f increases. Similarly, the intrusion length decreases with an increase in the tidal velocity ($\boldsymbol{F_d}^{-1}$ decreases with the second power of υ_0 and N^{-1} increases linearly with υ_0), which is understandable since the salt intrusion length at LWS is short if E is large.

Expression for the dispersion at the mouth and the salt intrusion length

Based on a large number of observations in real estuaries, a predictive expression for α_0^{HWS}, and hence for the dispersion D_0^{HWS} can be obtained. The observations made over the years in 13 estuaries, worldwide, are summarized in Table 5.3.

Similar to earlier researchers, a relation has been sought between non-dimensional parameters affecting dispersion such as: the Canter-Cremers number N, the densimetric Froude number $\boldsymbol{F_d}$, and the Estuarine Richardson number $N_R = N/\boldsymbol{F_d}$, defined earlier in Equation 2.36. The dispersion coefficient is made dimensionless by dividing it by the estuary depth and the tidal velocity amplitude. The following empirical relation was obtained by Savenije (1992a):

$$\frac{D_0^{HWS}}{\upsilon_0 h_0} = 1400\frac{E_0}{a}N_R^{0.5} \tag{5.70}$$

Apparently, the HWS dispersion coefficient at the mouth D_0^{HWS} varies with the root of the Estuarine Richardson number. Another prominent dimensionless ratio is E/a, which is the ratio of the horizontal mixing length to the convergence

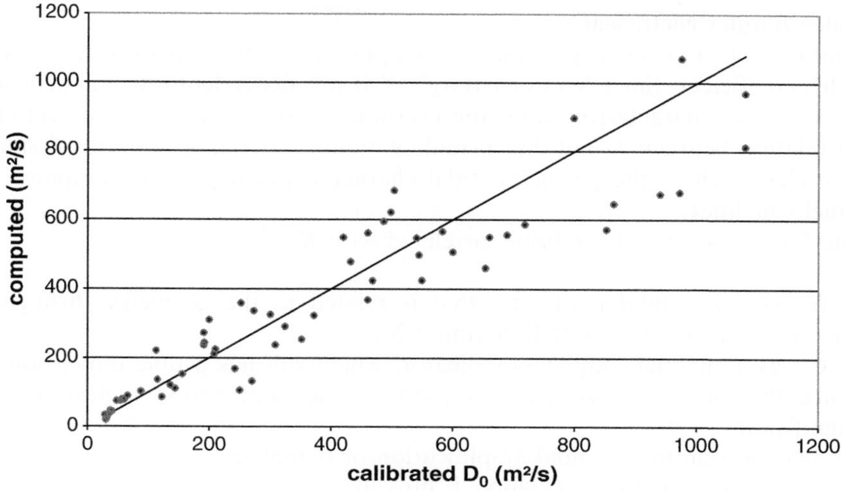

Figure 5.6 Empirical relation between values of D_0 at HWS computed with Equation 5.70 and values obtained from calibration.

length. Whereas the Estuarine Richardson number is a good indicator for (vertical) gravitational circulation, the ratio of horizontal mixing length to width convergence appears to be a good indicator for tide-driven mixing, particularly lateral residual circulation.

In Equation 5.70, a weighted value of a has been taken for those estuaries that demonstrate two branches in the function describing the longitudinal variation of the cross-sectional area. An average value of a is taken, weighted between a_1 and a_2 over the intrusion length. For these estuaries, the weighted value is shown in Table 5.3. It can be seen that the longer the salt intrusion length is, the closer a is to a_2.

A plot of Equation 5.70 against observations is presented in Figure 5.6. We can see that this very simple equation agrees surprisingly well with the observations, particularly taking into account the inaccuracies present in both the geometry and the hydrology. Figure 5.6 is a linear plot (not logarithmic, which tends to reduce the scatter). The line drawn in the middle is the line of perfect agreement, and it can be seen that this line fits the data points for a wide range of values. The R^2 correlation coefficient equals 0.88, with a standard error of 106 m^2/s, which is 30 percent of the average value of the observations. Due to the large uncertainty in the determination of the fresh water discharge, a large scatter in the data points is unavoidable. A relative error of 30 percent in the determination of D_0 could result in an error of 60 percent in Q_f. In estuaries, an error of 60 percent in determining the freshet may very well occur, particularly during low flow. Hence the relation obtained is quite satisfactory. The only way to improve the reliability of Equation 5.70 is by increasing the number of data points still further.

Van den Burgh's coefficient

It appears not to be so easy to find an adequate predictive equation for Van den Burgh's coefficient. Since Van den Burg's K is not dependent on time-dependent factors, but is a characteristic value for a certain estuary, it would be logical to look for a relation between K and dimensionless numbers defining the general state of the estuary, such as the geometry, tidal characteristics (e.g. tidal damping), and channel roughness.

The following ratios have been correlated with K:

1. E/H, as a key tidal parameter that is related to the geometry through the Geometry-Tide relation of Equation 2.90;
2. E/C^2, as a channel roughness indicator, which still has a time dimension, but since this time dimension is governed by the tidal period inclusion is not significant;
3. $(1-\delta b)$, accounting for tidal amplification or damping;
4. b/a, accounting for bottom slope, if present;
5. Ea/A'_0, as a ratio of tidal excursion to convergence: $E/(\mathrm{d}A/\mathrm{d}x)$;
6. H/h, as a key relation between tide and estuary shape.

The data for the regression analysis are presented in Table 5.4. The following relation has been obtained with a correlation coefficient $R^2=0.93$, which is reasonable but not always accurate. Surprisingly, the tidal range to depth ratio did not have a significant contribution in the multiple regression. This is correct

Table 5.4 Parameters used for the equation to predict Van den Burgh's coefficient

Estuary	K	T (s)	H_0 (m)	E_0 (km)	f	C^2 (m/s^2)	a (km)	b (km)	A_0' (km^2)	h (m)	δ (10^{-6}m^{-1})
Mae Klong	0.3	44,400	2	14	0.028	2809	102	155	1.4	5.20	−4.20
Lalang	0.65	86,400	3	31	0.023	3481	217	96	2.55	10.60	−1.00
Limpopo	0.5	44,440	1.1	8	0.026	3025	130	50	1.34	7.00	1.70
Tha Chin	0.35	86,400	2	15	0.039	2025	87	87	1.38	5.30	−9.40
Chao Phya	0.75	86,400	2.4	26	0.031	2500	109	109	4.3	7.20	−3.60
Incomati	0.15	44,440	1.4	7	0.022	3600	42	42	1.75	2.90	−13.0
Pungué	0.3	44,440	6.3	20	0.031	2500	21	21	26.4	3.50	−8.50
Maputo	0.38	44,440	2.8	13	0.022	3600	16	16	6.46	3.60	1.00
Thames	0.2	44,440	4.3	14	0.026	3025	23	23	58.5	7.10	2.30
Corantijn	0.21	44,440	2	11	0.026	3025	64	48	34.6	6.50	−1.70
Sinnamary	0.45	44,440	2.6	10	0.031	2500	39	39	1.21	3.80	−1.00
Gambia	0.6	44,440	1.2	10	0.031	2500	121	121	27.2	8.7	−1.00
Delaware	0.22	44,440	1.7	8	0.026	3025	41	42	255	6.60	1.70
Schelde	0.25	44,440	4	12	0.026	3025	28	28	150	10.50	3.80

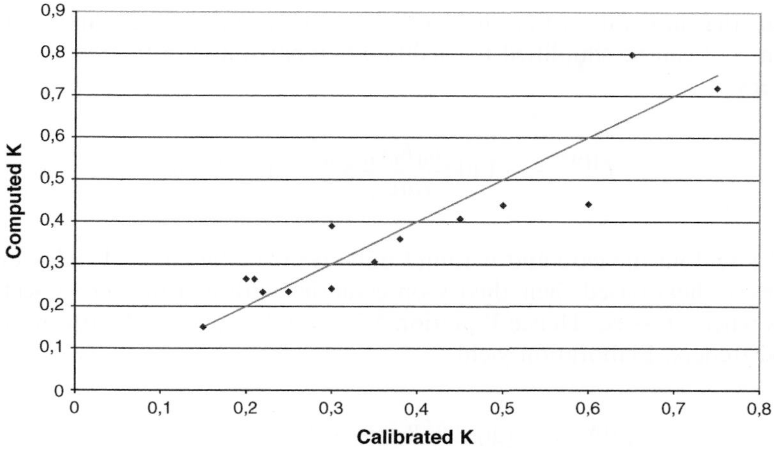

Figure 5.7 Van den Burgh's K computed by Equation 5.71 versus the values obtained from calibration.

because all the other parameters of the Geometry-Tide equation are already present in the regression analysis. We also see that tidal damping has a very strong impact on the value of K. Strongly damped estuaries with a high friction apparently have a small value of K. We can understand this if we consider that a high damping and roughness implies that a high amount of tidal energy is converted into mixing, which results into a relatively high importance of tidal mixing and relatively low importance of gravitational mixing.

Figure 5.7 presents the relation between calibrated values obtained in the various estuaries and the computed values with the following equation:

$$K = 0.2^{*}10^{-3}\left(\frac{E}{H}\right)^{0.65}\left(\frac{E}{C^2}\right)^{0.39}(1-\delta b)^{-2.0}\left(\frac{b}{a}\right)^{0.58}\left(\frac{Ea}{A_0'}\right)^{0.14} \tag{5.71}$$

We should keep in mind that the value of K should always be between zero and unity. The equation does not provide for that. We can see that K is particularly sensitive to tidal damping, channel roughness, and the tidal excursion. This equation should be used with caution. Its predictive value is weak. It should be used as a first estimate of K. Subsequently, a moving boat survey on a HWS or LWS situation will provide a more reliable value for K.

5.5.3 The predictive model compared to other methods

Combination of the empirical Equation 5.70 with Equations 5.47 and 5.48 yields:

$$L^{\mathrm{HWS}} = a\ln\left(-1400\frac{h_0 E_0 \upsilon_0}{Ka^2 U_0}N_{\mathrm{R}}^{0.5}+1\right) \tag{5.72}$$

Since on the interval $x(0,1)$: $\ln(x+1) \approx x$, as a first order approximation, Equation 5.72 can be simplified to facilitate comparison with the results of earlier researchers:

$$L^{\mathrm{HWS}} \approx -1400 \frac{h_0 E_0 \upsilon_0}{K a U_0} N_{\mathrm{R}}^{0.5} \quad \text{if:} \quad L < a \tag{5.73}$$

But this is seldom the case and is found only in prismatic channels. However, the earlier researchers based their theory on estuaries with constant cross section, or estuaries where $a \to \infty$. Hence Equation 5.73 may be compared with the work of these researchers. Elaboration yields:

$$L^{\mathrm{HWS}} \approx -1400 \frac{h_0 E_0}{K a} \boldsymbol{F}_{\mathbf{d}}^{-0.5} N^{-0.5} \quad \text{if:} \quad L < a \tag{5.74}$$

The correspondence of this equation with the relations obtained by Rigter (1973), Fischer (1974), and Van Os and Abraham (1990) presented in Equations 5.63, 5.66, and 5.69 is high, but it is highest with Van den Burgh's relation, Equation 5.68. The difference with Van den Burgh's, besides the logarithmic function of Equation 5.72, lies in the use of E/a. The tidal excursion is a very important longitudinal mixing length scale, and the convergence length a accounts for the lateral mixing through residual ebb and flood currents. If the estuary has a more pronounced funnel shape (a is small), this results in a large salt intrusion. Apparently the salt intrusion through lateral mixing is easier in a funnel-shaped estuary than in a prismatic estuary. This corresponds with what one would think intuitively. Hence the new formula takes better account of the topography, through the logarithmic function, and of the main mixing processes: N_{R} for gravitational circulation; E for longitudinal circulation through trapping; and a for lateral circulation between ebb and flood channels.

In Figure 5.8, the different predictive formulae for the salt intrusion length at HWS are compared for the estuaries studied. The data are presented in Table 5.5. It can be clearly seen that the new method predicts the salt intrusion length quite adequately, and much better than other methods.

Conclusion

The set of estuaries studied covers a wide range of dispersion coefficients as can be seen from Figure 5.6. The relation obtained is quite acceptable and is a considerable improvement over classical methods. The application of the steady state model in 15 estuaries with quite different tidal and geometrical characteristics has proven successful. The empirical relations for α and K turn the technique into a predictive method which can be used as a management tool to evaluate the effect of changes in the hydrology or the geometry of the estuary involved.

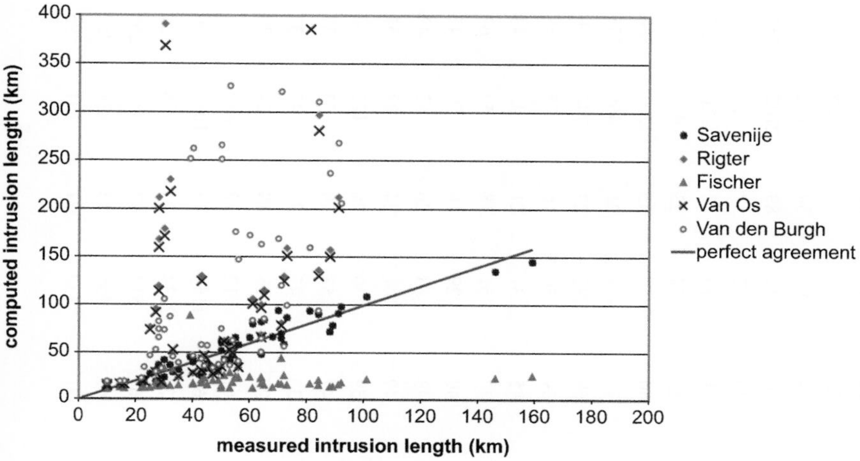

ure 5.8 Comparison of various predictive models for the salt intrusion length at HWS.

· UNSTEADY STATE MODEL

·.1 System response time

Section 5.4 expressions have been derived for the system response time as a ıction of the steady state salinity distribution $S(x)$. The first expression (T_K) is ;ed on equating the unsteady salinity variation to the variation of the salinity ween subsequent steady states as a result of fresh water flow depletion, which lded Equation 5.34. In this case, the system is assumed to be in a state of 90 ·cent equilibrium if $T_K < 0.1\ T_Q$, where T_Q is the time scale for the fresh water w depletion. The second method of deriving a system response time scale is using the time scale T_S needed to attain a new equilibrium condition, yielding uation 5.39. These equations are made dimensionless by comparing them to the shing time scale T_f: Equation 5.41. Both T_S and T_K require an expression for ;). After the derivations carried out in Section 5.5, this expression is available.

Savenije (1992a), taking $X = L/2$ performed the integration of these equations integer values of $1/K$, which are not repeated here. If $1/K$ is not an integer mber, the equations can only be solved numerically. As an example the ılytical equations for $K = 0.5$ are presented:

$$T_K = -\frac{2\beta a A_0}{Q_f} \frac{\frac{L}{2a}(1+2\beta) - 1}{\left(1 - \beta\left(\exp\left(\frac{L}{2a}\right) - 1\right)\right)^2} \quad \text{if:} \quad K = 0.5 \tag{5.75}$$

$$T_S = -\frac{\beta(1+\beta) a A_0}{Q_f} \frac{\exp\left(\frac{L}{2a}\right) - \exp\left(-\frac{L}{2a}\right) - \frac{L}{a}}{\left(1 - \beta\left(\exp\left(\frac{L}{2a}\right) - 1\right)\right)^2} \quad \text{if:} \quad K = 0.5 \tag{5.76}$$

Table 5.5 Measured and predicted salt intrusion length in different estuaries

Estuary	Date	H_0 (m)	E_0 (km)	Q_f (m^3/s)	f	A_0 (km^2)	h (m)	N	F^2	F_d	L_{HWS} (km)	Savenije (km)	Rigter (km)	Fischer (km)	V.Os (km)	Burgh (km)
Mae Klong	20/01/77	2.0	11	120	0.028	1.4	5.2	0.67	0.00	0.16	29	22	18	15	19	34
	08/03/77	1.5	10	60	0.028	1.4	5.2	0.19	0.01	0.46	23	22	19	12	19	35
	09/04/77	2.0	14	36	0.028	1.4	5.2	0.08	0.02	0.93	35	32	24	16	25	40
Solo	26/07/88	0.8	9	50	0.023	2.07	9.2	0.23	0.00	0.05	30	24	180	34	172	74
	08/09/88	0.4	5	7	0.023	2.07	9.2	0.06	0.00	0.01	39	46	2242	89	2102	252
Lalang	20/10/89	2.6	27	120	0.023	2.55	10.6	0.15	0.01	0.52	33	30	52	32	53	46
Limpopo	04/04/80	1.1	7	150	0.026	1.71	7	0.56	0.00	0.17	22	21	19	12	20	27
	31/12/82	1.1	8	2	0.026	1.71	7	0.01	0.00	0.19	64	83	1054	23	989	196
	22/04/83	0.5	4	1	0.026	1.71	7	0.01	0.00	0.05	71	80	3953	45	3703	386
	24/07/94	0.9	6.8	2	0.026	1.71	7	0.02	0.00	0.14	55	57	498	21	468	135
	10/08/94	1.0	7.1	2	0.026	1.71	7	0.01	0.00	0.15	60	69	792	23	744	170
Tha Chin	16/04/81	1.6	12	55	0.039	3	5.3	0.13	0.00	0.20	45	39	36	16	35	58
	27/02/86	2.6	20	40	0.039	3	5.3	0.03	0.04	1.74	43	45	31	21	32	44
	01/03/86	1.8	14	40	0.039	3	5.3	0.08	0.00	0.21	54	47	51	18	50	63
	13/08/87	2.0	15	39	0.039	3	5.3	0.04	0.02	1.12	40	40	29	16	29	47
Chao Phya	05/06/62	2.2	22	63	0.031	5.3	7.2	0.06	0.01	0.44	51	43	63	26	62	43
	17/03/80	1.5	18	43	0.031	5.3	7.2	0.05	0.01	0.27	64	49	101	24	97	51
	28/03/80	1.7	20	31	0.031	5.3	7.2	0.03	0.01	0.31	72	59	130	26	125	57
	29/01/83	2.4	26	90	0.031	5.3	7.2	0.07	0.01	0.54	53	42	53	29	53	37
	23/02/83	1.6	19	100	0.031	5.3	7.2	0.11	0.01	0.34	44	33	47	23	47	37
	16/01/87	2.5	15	180	0.031	5.3	7.2	0.24	0.00	0.29	27	20	28	19	29	30
Incomati	30/07/80	1.4	7	3	0.022	8.1	2.9	0.00	0.01	0.38	53	61	712	12	668	327
	05/09/82	1.4	7	2	0.022	8.1	2.9	0.00	0.01	0.34	68	67	1164	13	1092	400
	10/02/83	1.2	8	1	0.022	8.1	2.9	0.00	0.01	0.45	89	79	2034	14	1906	529
	23/06/93	1.4	8	4	0.022	8.1	2.9	0.00	0.01	0.45	50	61	514	12	482	266
	07/07/93	2.6	9	4	0.022	8.1	2.9	0.00	0.01	0.57	50	63	458	13	431	252
Pungué	26/09/80	6.3	20	22	0.031	26.4	3.5	0.00	0.06	2.40	84	91	137	21	131	95

	26/05/82	5.0	10	50	0.031	26.4	3.5	0.01	0.01	0.64	61	80	107	12	102	84
	06/08/82	5.2	14	36	0.031	26.4	3.5	0.00	0.03	1.17	65	85	116	16	110	86
	22/09/82	5.2	14	26	0.031	26.4	3.5	0.00	0.03	1.14	73	87	160	16	152	100
	29/10/82	6.0	16	60	0.031	26.4	3.5	0.01	0.04	1.53	64	83	69	17	67	65
	03/10/93	5.3	15	10	0.031	26.4	3.5	0.00	0.03	1.31	81	94	371	17	349	153
	12/10/93	3.8	15	10	0.031	26.4	3.5	0.00	0.03	1.29	70	93	376	17	353	153
	16/10/93	6.4	16	10	0.031	26.4	3.5	0.00	0.04	1.47	92	95	354	18	333	148
	31/01/02	6.2	20	262	0.031	26.4	3.5	0.02	0.06	2.91	47	35	27	21	28	35
	27/02/02	6.1	20	200	0.031	26.4	3.5	0.02	0.06	2.81	49	37	30	21	30	38
	01/03/02	6.7	24	150	0.031	26.4	3.5	0.01	0.08	5.19	56	40	35	25	35	42
Maputo	28/04/82	2.8	13	25	0.022	40	3.6	0.00	0.02	0.96	30	43	391	16	369	106
	15/07/82	1.5	6	8	0.022	40	3.6	0.00	0.01	0.20	40	44	2576	18	2413	263
	06/04/82	2.9	10	180	0.022	40	3.6	0.02	0.01	0.57	25	28	77	13	74	47
	19/04/82	3.3	12	120	0.022	40	3.6	0.01	0.02	0.82	27	31	96	14	92	53
	02/05/84	3.4	13	70	0.022	40	3.6	0.01	0.02	1.20	28	34	120	15	115	66
	17/05/84	3.3	14	50	0.022	40	3.6	0.00	0.03	1.25	28	37	169	16	160	75
	29/05/84	2.8	12	40	0.022	40	3.6	0.00	0.02	0.95	32	37	231	15	218	88
	02/08/84	2.8	11	49	0.022	40	3.6	0.00	0.02	0.77	28	36	212	14	201	83
Thames	07/04/49	5.3	14	40	0.026	58.5	7.1	0.00	0.01	0.60	101	109	1005	22	944	491
Corantijn	02/07/65	2.0	11	1995	0.026	69	6.5	0.12	0.01	0.38	50	53	36	15	36	76
	16/08/65	2.2	12	680	0.026	69	6.5	0.04	0.01	0.45	71	71	81	17	79	121
	08/12/78	1.8	10	115	0.026	69	6.5	0.01	0.01	0.55	84	92	298	16	281	311
	14/12/78	2.3	12	130	0.026	69	6.5	0.01	0.01	0.83	91	92	213	16	202	269
	20/12/78	1.6	9	220	0.026	69	6.5	0.02	0.01	0.49	88	73	158	14	151	238
Sinnamary	12/11/93	2.6	10	148	0.031	3.5	3.8	0.19	0.01	0.71	10	12	13	11	14	18
	27/04/94	2.9	12	108	0.031	3.5	3.8	0.11	0.02	1.19	10	13	15	13	16	20
	02/11/94	2.7	11	106	0.031	3.5	3.8	0.12	0.02	0.78	16	13	16	12	17	20
	03/11/94	2.9	12	106	0.031	3.5	3.8	0.11	0.02	1.02	15.5	13	16	13	17	20
Delaware	23/08/32	1.7	8	120	0.026	255	6.6	0.00	0.00	0.22	146	136	2122	24	1989	633
	04/10/32	1.7	8	72	0.026	255	6.6	0.00	0.00	0.22	159	146	3533	26	3310	816

As L/a can be written as a function of β, using Equation 5.48, these are functions of β and Q_f. Since β is also essentially a function of Q_f, because D_0, is a function of Q_f, the time scales are primarily driven by the freshet. Also the expression for T_f of Equation 5.41 can be written as a function of β.

$$T_f = -\frac{A_0 a}{Q_f}\left(1 - \exp\left(-\frac{L}{a}\right)\right) = -\frac{A_0 a}{Q_f}\frac{1}{(1+\beta)} \tag{5.77}$$

Combination leads to the dimensionless time-scales:

$$\frac{T_K}{T_f} = 2\beta(1+\beta)\frac{\frac{L}{2a}(1+2\beta) - 1}{\left(1 - \beta\left(\exp\left(\frac{L}{2a}\right) - 1\right)\right)^2} \quad \text{if:} \quad K = 0.5 \tag{5.78}$$

$$\frac{T_S}{T_f} = \beta(1+\beta)^2\frac{\exp\left(\frac{L}{2a}\right) - \exp\left(-\frac{L}{2a}\right) - \frac{L}{a}}{\left(1 - \beta\left(\exp\left(\frac{L}{2a}\right) - 1\right)\right)^2} \quad \text{if:} \quad K = 0.5 \tag{5.79}$$

Because L/a is a sole function of β, these equations can be written as sole functions of β as well. In Table 5.6, the values of T_S, T_K, and T_f have been presented and compared, for the various estuaries under a minimum flow Q_0, together with values of T_Q, where available. The values of T_K and T_S indicate the same pattern. It can be seen that in several cases $T_K < T_Q$ (Pungué, Maputo, Corantijn, Sinnamary and Delaware) but that in other cases $T_K > T_Q$ (Limpopo, Incomati, Thames and Gambia). The value of T_S is larger than T_Q only in the Gambia, and in the Incomati they are equal. Of these estuaries, the Gambia is clearly in unsteady state, the Incomati, the Limpopo and Thames are on the limit. For the Limpopo and Incomati, unsteady state occurs only during the lowest minimum flow. Since both the values of T_K and T_S are inversely proportional to the root of Q_f, a modest increase in the discharge would bring about equilibrium (see Savenije 1992a). In applying the steady state model to the lowest flow situation, where equilibrium is not completely reached, we obtain a conservative estimate of the actual salt intrusion length, which is not so bad. Moreover, the Thames, the Limpopo, and the Incomati have minimum flows which are regulated as a result of reservoir release and upstream withdrawal. Therefore, the time scale of regulated flow is much longer during minimum flow than the time scale of the natural recession curve. So the use of the steady state model in these estuaries is acceptable, even during low flow.

Hence, of all estuaries studied, the only estuary where the steady state model cannot be applied is the Gambia. This estuary will be used to demonstrate the unsteady state model.

Table 5.6 System response time in the dry season, in relation to water particle travel time T_f and hydrological time scale

Estuaries	K	h (m)	a (km)	A_0' (m^2)	Q_0 (m^3/s)	L (km)	L/a	β	T_f (days)	T_K/T_f (days)	T_S/T_f (days)	T_K (days)	T_S (days)	T_Q (days)
Mae Klong	0.30	5.2	102	1400	−30	26	0.25	3.44	12	0.55	0.12	7	1	reg
Solo	0.60	9.2	226	2070	−10	35	0.15	5.97	78	0.63	0.19	49	15	reg
Lalang	0.65	10.6	217	2550	−50	65	0.30	2.86	33	0.58	0.19	19	6	
Limpopo	0.50	7.0	130	1400	−5	60	0.46	1.70	156	0.53	0.16	82	26	45
Tha Chin	0.35	5.3	87	1380	−10	70	0.80	0.81	77	0.43	0.13	33	10	
Chao Phya	0.75	7.2	109	4300	−30	50	0.46	1.72	67	0.53	0.21	35	14	
Incomati	0.15	3.0	42	1520	−1	50	1.19	0.44	514	0.36	0.07	185	36	36
Pungué	0.30	5.3	20	28000	−20	70	3.50	0.03	314	0.09	0.08	28	25	226
Maputo	0.38	3.6	16	6460	−10	40	2.50	0.09	110	0.16	0.11	18	12	103
Thames	0.20	7.1	23	58500	−20	90	3.91	0.02	763	0.08	0.06	61	46	50
Corantijn	0.21	6.5	64	34600	−500	50	0.78	0.84	28	0.43	0.09	12	3	58
Sinnamary	0.45	10.0	35	4000	−100	16	0.46	1.73	6	0.60	0.19	4	1	
Gambia	0.60	8.7	121	27200	−2	300	2.48	0.09	17450	0.15	0.13	2618	2269	42
Schelde	0.25	10.0	26	150000	−90	110	5.23	0.01	494	0.06	0.06	30	30	124
Delaware	0.22	6.6	41	255000	−300	140	3.41	0.03	390	0.1	0.07	39	27	101

5.6.2 Unsteady state dispersion

It has been mentioned that the steady state dispersion coefficient D_{SS} is not necessarily the same as the unsteady state dispersion coefficient D. This is particularly important in estuaries where the system lags considerably behind the steady state situation, as is the case in the Gambia.

It should be borne in mind that none of the expressions in use for the dispersion is completely correct, physically; even the one-dimensional dispersion equation itself lacks a full physical basis. All equations in use for $D(x)$ are (at least partially) empirical, whether they are considered constant, a function of $\partial S/\partial x$ (Thatcher and Harleman, 1972), or a function of $(\partial S/\partial x)^2$ (Chatwin and Allen, 1985). In Chapter 4, it has been shown that the method where D is proportional to S^K can safely be added to this list and that it has a wide range of applicability.

Depending on which method is used to determine the dispersion, the unsteady state model will react differently. If the dispersion is computed on the basis of $\partial S/\partial x$, $(\partial S/\partial x)^2$ or S^K, then, in a state of disequilibrium (unsteady state), the dispersion is different from the steady state dispersion, $D \neq D_{SS}$, simply because the salinity distribution is different. Such a method is a 'status quo' method; it uses the dispersion that corresponds to the present salinity distribution, irrespective of whether the system is in equilibrium or not. Here the 'status quo' equation applied to the unsteady state would read:

$$\frac{D}{D_0} = \left(\frac{S}{S_0}\right)^K \tag{5.80}$$

If however, a steady state model for the dispersion is used, e.g. Van den Burgh's method where the gradient of the steady state dispersion is inversely proportional to the cross-sectional area, then the outcome is different. The use of the steady state equation leads to a simulation where the dispersion coefficient used corresponds with the ultimate state of equilibrium that would occur if the discharge were maintained constant over a sufficiently long period. The steady state dispersion reads:

$$\frac{D_{SS}}{D_0} = 1 - \beta\left(\exp\left(\frac{x}{a}\right) - 1\right) \tag{5.81}$$

If we use a steady state dispersion model to a situation of unsteady state, we implicitly assume that the main factors describing the dispersion are time-invariant (e.g. the geometry and key hydraulic parameters, such as channel roughness, the ratio E/H, the phase lag ε, or the tidal damping), or that (unlike the salinity itself) the dispersion reacts directly to a change in the river discharge (much like the mass balance which affects buoyancy). Particularly, if dispersion is mostly driven by residual circulation, which depends on tidal characteristics and geometry, then this would be an acceptable approach. If, however, the dispersion is mostly density-driven, depending on the salinity gradient, then the application of Equation 5.80 to the unsteady state would be best. In reality, dispersion is a combination of both mechanisms, and hence a combination of Equations 5.80 and 5.81 appears to be the best approach.

Consider the hypothetical case where the fresh water discharge is suddenly decreased from Q_0 to Q_1 and where the salinity distribution needs considerable time to adjust to the new situation. A 'status quo' method which computes the dispersion on the basis of the instantaneous longitudinal salinity distribution would then, for some time, continue to use a dispersion coefficient that corresponds essentially to a situation where $Q = Q_0$. The 'ultimate equilibrium' method, however, would immediately after the decrease in the discharge start to use a dispersion that corresponds with $Q = Q_1$, which at the toe of the intrusion curve leads to more dispersion. Hence, the 'status quo' method reacts considerably slower than the 'ultimate equilibrium' method in adjusting itself to the new situation.

Application of the 'status quo' Equation 5.80, has the disadvantage that dispersion is blocked at the toe of the salt intrusion curve. Application of the 'ultimate equilibrium' method, however, has the disadvantage of too much dispersion at the toe of the salt intrusion curve. Which of the two approaches is the best is difficult to say, and moreover, irrelevant as neither of the methods is physically completely 'correct.'

The method followed here is a practical one based on experiences gained with the Gambia estuary. The best result was obtained with an intermediate method that interpolates between the 'status quo' dispersion and the 'ultimate equilibrium' dispersion. With this method the model gradually converges to the steady state situation. A weighting factor is used as a calibration coefficient. In the case of the Gambia, equal weights were given to the 'status quo' and 'ultimate equilibrium' dispersion.

A special situation occurs when the fresh water discharge at the estuary mouth becomes negative, as may be the case when the net evaporation from the estuary surface exceeds the fresh water inflow into the estuary. The equation for the boundary condition D_0^{HWS}, Equation 5.70, then no longer applies. For that situation, the dispersion of the 'ultimate equilibrium' is assumed to be constant. The value that best fits the measurements is found through calibration. This special situation is only important in estuaries where evaporation plays an important role, as is the case in the estuaries discussed in the following sections: the Gambia, the Saloum, and the Casamance.

5.6.3 Application of the unsteady state model

An unsteady state equation can be solved using a six-point implicit finite difference scheme as suggested by Fischer et al. (1979). The unsteady state model makes use of the unsteady state equation that involves rainfall and evaporation,

Equation 5.24, in combination with Van den Burgh's equation, Equation 5.43. A combination of these equations yields:

$$r_S\frac{\partial s}{\partial t}+(1-K)\frac{Q_r}{A}\frac{\partial s}{\partial x}-(1-K)\frac{rb}{\bar{h}}\frac{\partial s}{\partial x}+\frac{D}{a}\frac{\partial s}{\partial x}-D\frac{\partial^2 s}{\partial x^2}+s\frac{r}{h_0}=0 \qquad (5.82)$$

This equation can be written as:

$$r_S\frac{\partial s}{\partial t}+q\frac{\partial s}{\partial x}-D\frac{\partial^2 s}{\partial x^2}+\frac{r}{h_0}s=0 \qquad (5.83)$$

with:

$$q=(1-K)\frac{Q_r-Brb}{A}+\frac{D}{a} \qquad (5.84)$$

In the six-point finite differences scheme, this equation is for each time-step converted into a tri-diagonal matrix of x-dependent coefficients, which can be solved by a Gaussian elimination method, as described by Carnahan et al. (1969, pp. 440–442). The details are described in Savenije (1992a).

The model functions very efficiently and has been successfully applied in the Gambia (Savenije, 1988; Risley et al., 1993), the Schelde, the Saloum, and the Casamance (Savenije and Pagès, 1992). The application in the Gambia is presented below. The application in the Saloum and Casamance are presented in Section 5.7.

Application to the Gambia estuary

Table 5.1 indicates that evaporation is important in the salinity distribution along the Gambia estuary. The question however is how important the effect is in quantitative terms. In Figure 5.9a, the calibration of the unsteady state model without the influence of rainfall and evaporation, is shown against the longitudinal salinity distribution measured (thick lines) at different times during the hydrological year 1972/73 (Savenije, 1988). It should be observed that the measurements have been carried out somewhat haphazardly, not taking into account the time of day (HWS, LWS, or TA) which may lead to a position error of 5 km. However, it is clear from Figure 5.9a that the maximum intrusion at the end of May in the central part of the salt intrusion curve is not reached, and that the minimum intrusion reached by the model at the end of September is too high.

The position of the toe of the salt intrusion curve (the point where $s \approx S_f$), however, is correctly modelled, which is illustrated by Figure 5.10a. This figure follows the location of the toe of the salt intrusion curves ($s = 1\,\text{kg/m}^3$) with time. The problem faced, at that time, was to improve the fit of the longitudinal distribution, while not affecting the total salt intrusion length. The solution to this

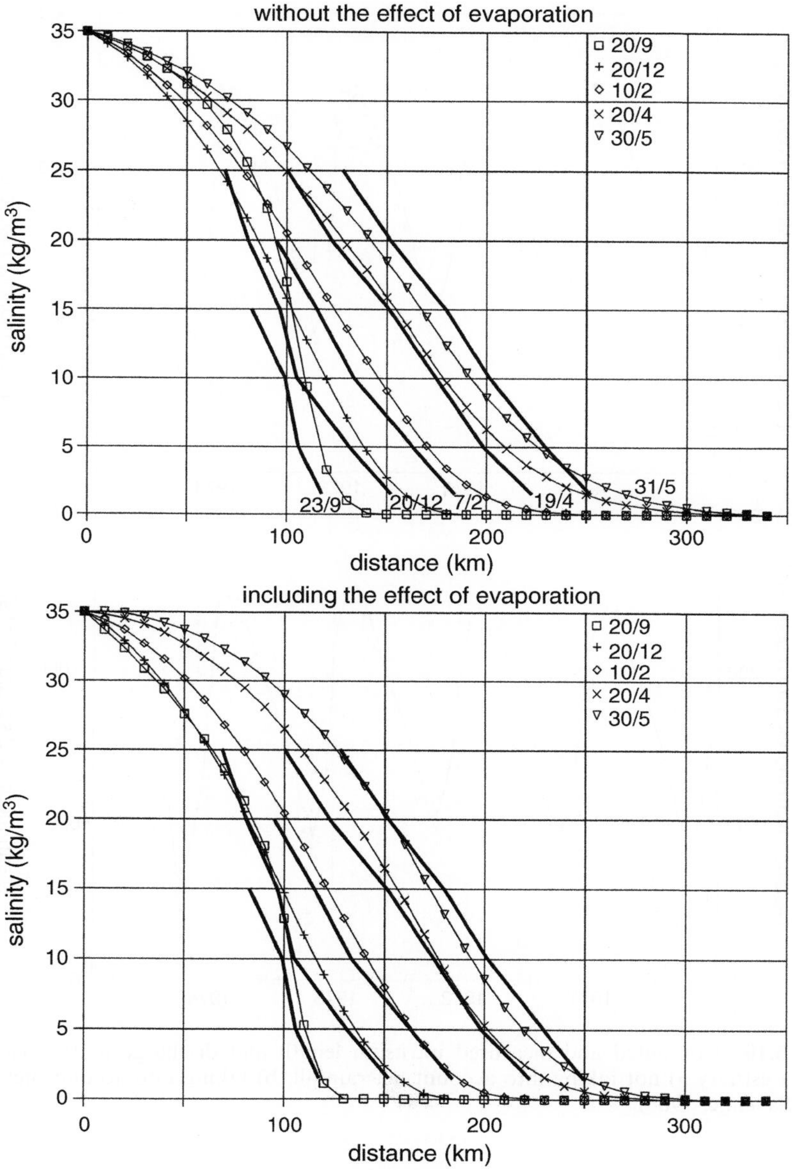

Figure 5.9 Computed and measured longitudinal salt intrusion distribution in the Gambia estuary, a) not taking into account net rainfall, b) taking into account net rainfall (after: Savenije, 1988).

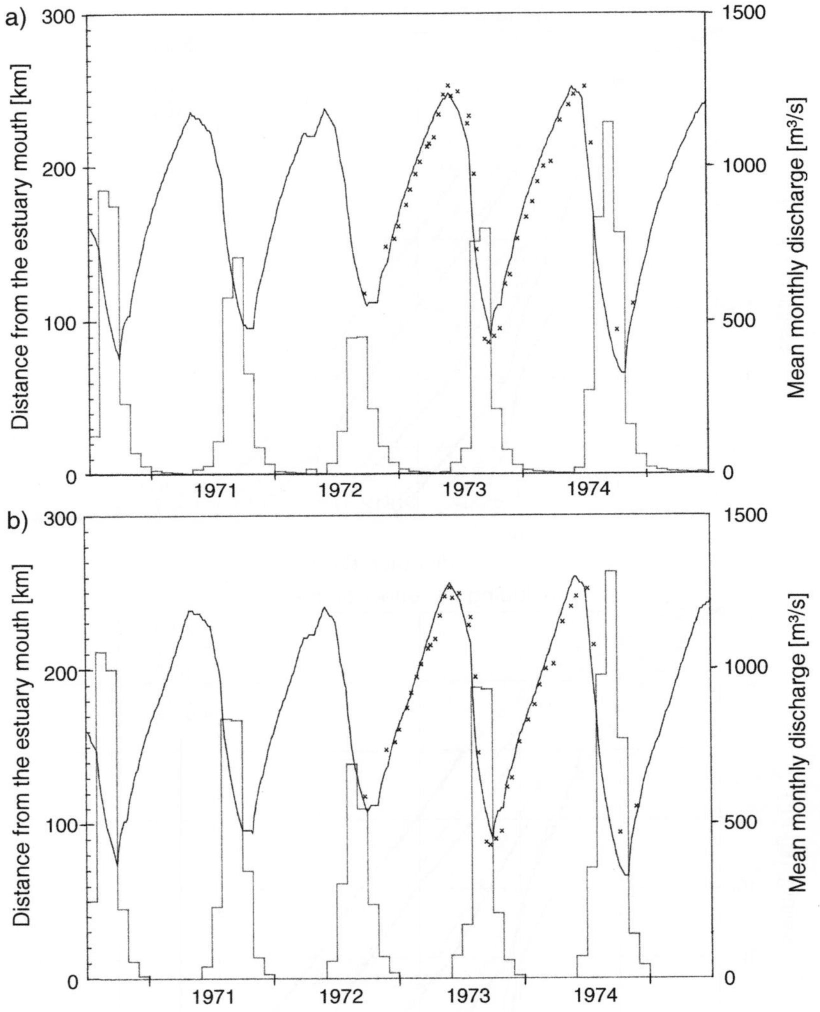

Figure 5.10 Computed and measured intrusion length and discharge to the sea in The Gambia estuary a) not taking into account net rainfall; b) taking into account net rainfall (after: Savenije, 1988).

problem was the incorporation of r, the net rainfall, which is the difference between rainfall and evaporation. The toe of the curve is not directly affected by r either through the third term of Equation 5.25, since $\partial s/\partial x = 0$ at the toe, or through the sixth term, since $s = 0$ at the toe. It is only indirectly affected by an increase of the salinity in the central part of the estuary, which, through dispersion, propagates upstream. Figure 5.10b shows the position of the toe of the salt intrusion if the effect of rainfall and evaporation is taken into account. Figures 5.10a and 5.10b are essentially the same. The only apparent difference between the figures lies in the discharge of fresh water to the ocean.

Figure 5.9b shows the longitudinal distribution after inclusion of the rainfall terms. The fit, although not perfect, is certainly better than in Figure 5.9a. It can be seen that the upstream and downstream limits of the intrusion are essentially at the same position, but that the curves have become more concave in the dry season, as a result of excess evaporation, and less concave in the wet season, as a result of excess rainfall. The difference in runoff at the estuary mouth is considerable as a result of rainfall and evaporation, as can be concluded from comparing Figures 5.10a and 5.10b.

The notion that evaporation played an important role in the Gambia estuary was strongly supported by Pagès and Citeau (1990) who stated that the Sahelian estuaries: Gambia, Sénégal, Casamance and Saloum, in that order, were all turning more and more saline due to the drought of the 1980s. In the Sénégal, the salinity was affected in much the same way as in the Gambia, but the Saloum and Casamance functioned as hypersaline estuaries.

5.7 HYPERSALINE ESTUARIES

An estuary may become hypersaline if the salt flux F is not sufficient to evacuate the salt accumulation resulting from evaporation. The fresh water discharge of the Gambia is still too large for the estuary to become hypersaline, but the two estuaries bordering the Gambia to the North and to the South, the Saloum and the Casamance, are strongly hypersaline.

The Saloum has always been hypersaline. Measurements in the Saloum near a salt production farm go as far back as 1965 (Pagès and Citeau, 1990). The Casamance, however, although strongly influenced by evaporation, has only become hypersaline during the Sahelian drought, which started in the late 1970s. Around 1980, an ecological disaster took place in the Casamance. The fresh and brackish habitats turned hypersaline (up to 100 kg/m^3), blocking off migration routes for migrant species and stunting the growth of otherwise salt-tolerant vegetation (Savenije and Pagès, 1992).

Figure 5.11 shows model results of the Saloum against measurements at three locations along the estuary. It can be seen that pronounced hypersaline conditions have always existed due to the ephemeral character of the rivers entering the estuary. The fit of the model is not perfect. This is largely due to the lack of data on fresh water discharge into the estuary. A simple hydrological model had to be made on the basis of rainfall data to simulate inflow series. Moreover, hydrographical data on cross section and depth were scarce. Since the effect of evaporation on salt accumulation, to a large extent, depends on the depth (see Equation 5.25), this lack of information strongly limits the accuracy of the model. In qualitative terms, however, the model is quite reliable, as can be judged

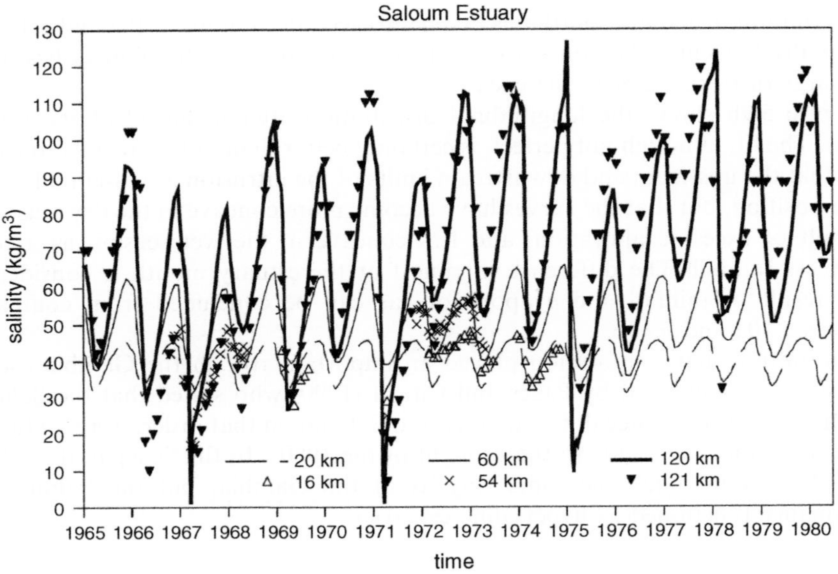

Figure 5.11 Computed and measured salinity variation along the Saloum estuary (after: Savenije and Pagès, 1992).

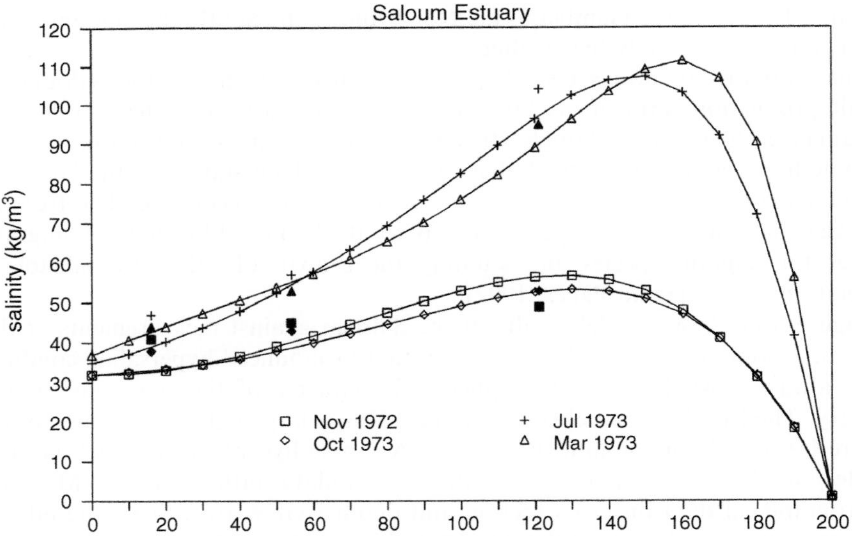

Figure 5.12 Computed and measured longitudinal distribution of the salinity along the Saloum estuary (after: Savenije and Pagès, 1992).

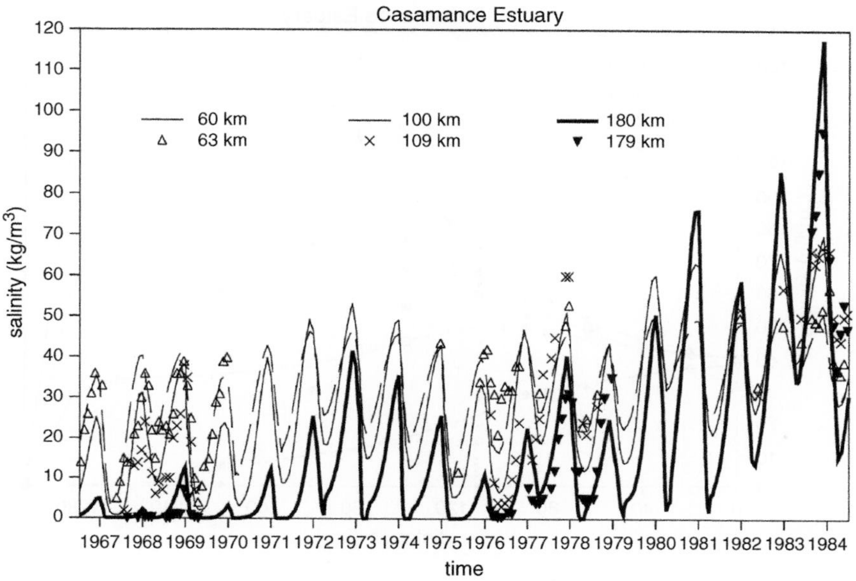

Figure 5.13 Computed and measured salinity variation along the Casamance estuary (after: Savenije and Pagès, 1992).

from the longitudinal profiles presented in Figure 5.12, where a comparison is made between measured and modelled salinities.

Figures 5.13 and 5.14 show similar graphs for the Casamance estuary. In the casamance, it can indeed be seen that since the start of the Sahelian drought the environment has completely changed from a normal estuary into a hyper-saline estuary. Until the year 1981, the estuary, at a distance of 180 km from the mouth, turned fresh annually. After that time, serious salinization took place, which is dramatically illustrated by Figure 5.14 in the longitudinal profile. The difference between the situations of 1978 and 1984 is striking.

Although data on fresh water inflow were available in the Casamance, the lack of reliable data on depths and cross sections seriously hampered the calibration. Nevertheless, it appeared possible to apply the methodology described in this study to an extreme situation for which it had not been developed originally. Given the limited amount of data available, the model performs well.

5.8 CONCLUDING REMARKS

In this chapter, a predictive model for salt intrusion in well-mixed alluvial estuaries has been presented. In most cases the equation for steady state can be applied to the HWS situation, yielding a very simple equation to predict the intrusion length: Equation 5.72. This equation is physically based in that it relates to the

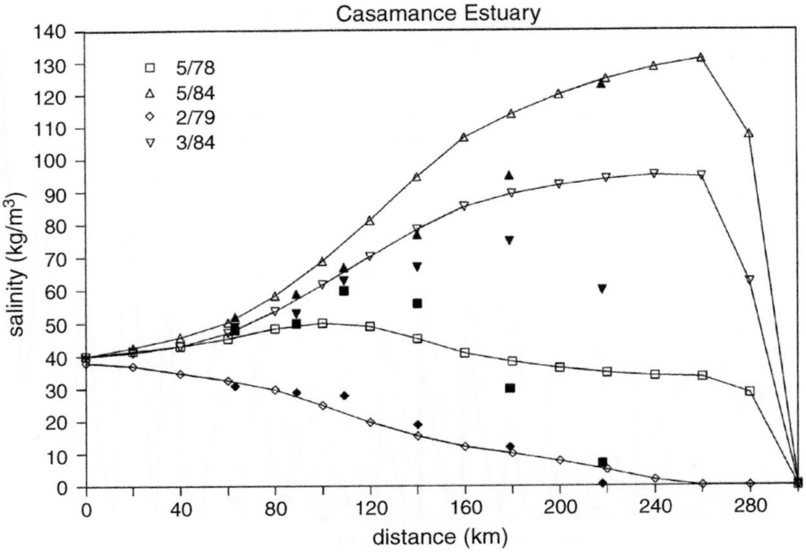

Figure 5.14 Computed and measured longitudinal salinity distribution along the Casamance estuary (after: Savenije and Pagès, 1992).

main driving mechanisms for salt intrusion: the gravitation circulation (determined by the Estuarine Richardson Number), and the residual circulation (determined mainly by E/a). Also in the unsteady state situation the method provides a useful tool, which is demonstrated by application in the Gambia and in hypersaline estuaries, such as the Saloum and the Casamance.

The strength of the method lies in its simplicity, while retaining its physical basis. The disadvantage of the method is that it requires a simplified topography. In cases where the topography is complex, one may have to rely on a two-dimensional model. But such models require large amounts of data. Even in case a more complex model is required, the one-dimensional model presented here can provide valuable information for the organization of a hydrometric survey: with the one-dimensional model, one can see which variables are most important and in which, density of observation.

6

References

1. Abbott, M.R. (1960). Boundary layer effect in estuaries. *Journal of Marine Research*, **18**(2), 83–100.
2. Airy, G.B. (1845). Tides and waves. In *Encyclopaedia Metropolitana*, Vol. 5, London, pp. 241–396.
3. Beven, K. (1993). Prophesy, reality and uncertainty in distributed hydrological modelling. *Advances in Water Resources*, **16**, 41–51.
4. Blench, T. (1952). Regime theory for self-formed sediment-bearing channels. *Transactions ASCE*, **117**, 383–408.
5. Bowden, K.F. (1967). Circulation and diffusion. In *Estuaries: Physical Factors* (G.H. Lauff, ed.), AAAS Publication No. 85, Washington D.C., pp.15–36.
6. Bowden, K.F. (1981). Turbulent Mixing in Estuaries. *Ocean Management*, **6**, 117–135.
7. Bretting, A.E. (1958). Stable channels. *Acta Polytechnica Scandinavia 245*, Copenhagen, Denmark. Moolers Bogtrykkeri.
8. Bruun, P. and Gerritsen, F. (1960). *Stability of Coastal Inlets.* North Holland, Amsterdam.
9. Canter Cremers, J.J. (1921). Eenige beschouwingen over de waterbeweging en de beweging van vaste stoffen in benedenrivieren, getoetst aan uitkomsten van waarnemingen en aan de uitwerking van uitgevoerde berekeningen in den Waterweg van Rotterdam naar zee. *De ingenieur*, **29**.
10. Carnahan, B. Luther, H.A. and Wilkes, J.O. (1969). *Applied Numerical Methods.* John Wiley & Sons, Inc., New York.
11. Chatwin, P.C. and Allen, C.M. (1985). Mathematical models of dispersion in rivers and estuaries. *Annual Review of Fluid Mechanics*, **17**, 119–149.
12. D'Alpaos, A. Lanzoni, S. Marani, M. Fagherazzi, S. and Rinaldo, A. (2005). Tidal network ontogeny: channel initiation and early development. *Journal of Geophysical Research*, **110**. F02001 1–14. F2 april doi: 10.1029/2004JF000182.
13. Davies, L.J. (1964). A morphogenic approach to the worlds' shorelines. *Zeitschrift für Geomorphologie*, **8**, 127–142.
14. Dronkers, J.J. (1964). *Tidal Computations in Rivers and Coastal Waters.* North-Holland Publishing Company, Amsterdam.
15. Dronkers, J. (1982). Conditions gradient-type dispersive transport in one-dimensional, tidally averaged transport models. *Estuarine, Coastal and Shelf Science*, **14**, 599–621.
16. Dronkers, J. Van Os and Leendertse. (1981). Predictive Salinity Modelling of the Oosterschelde with Hydraulic and Mathematical Models. In *Transport models for inland and coastal waters.* Academic Press Inc. London, pp. 451–482.
17. Dyer, K.R. (1973). *Estuaries: a physical introduction.* John Wiley & Sons, Aberdeen, UK.
18. Dyer, K.R. (1974). The salt balance in stratified estuaries. *Estuarine, Coastal and Shelf Science*, **2**, 273–281.

19. Dyer, K.R. (1997). *Estuaries: a physical introduction* (new revised issue). John Wiley & Sons, Aberdeen, UK.
20. Engelund, F. and Hansen, E. (1967). *A Monograph on Sediment Transport in Alluvial Streams.* Teknisk Vorlag, Copenhagen, Denmark.
21. Fischer, H.B. (1972). Mass transport mechanisms in partially mixed estuaries. *Journal of Fluid Mech*anics, **53**(4), 671–687.
22. Fischer, H.B. (1974). Discussion of 'Minimum length of salt intrusion in estuaries' by B.P. Rigter, 1973. *Journal of the Hydraulics Division Proceedings*, **100**, 708–712.
23. Fischer, H.B. List, E.J. Koh, R.C.Y. Imberger J. and Brooks, N.H. (1979). *Mixing in inland and coastal waters.* Academic Press, New York.
24. Friedrichs, C.T. and Aubrey, D.G. (1994). Tidal propagation in strongly convergent channels, *Journal of Geophysical Research*, **99**(C2), 3321–3336.
25. Graas, S. (2001). *Verloop van het Getijverschil over het Schelde-estuarium.* M.Sc. Thesis, TU Delft, The Netherlands.
26. Green, G. (1837). On the motion of waves in a variable canal of small depth and width. *Transactions of Cambridge Philosophical Society*, **6**, 457–462.
27. Hansen, D.V. (1965). Currents and mixing in the Columbia River estuary. Ocean Science and Ocean Engineering. *Transaction of the Joint Conference of the Marine Technological Society and American Society of Limnology and Oceanography*, 943–955.
28. Hansen, D.V. and Rattray, M. (1965). Gravitational circulation in straits and estuaries. *Journal of Marine Research*, **23**, 104–122.
29. Harleman, D.R.F. (1966). Tidal dynamics in estuaries Part II: Real estuaries. In *Estuary and Coastline Hydrodynamics*, (A.T. Ippen, ed.), (McGraw-Hill, New York, USA, 522–545.
30. Harleman, D.R.F. and Thatcher, M.L. (1974). *Longitudinal Dispersion and Unsteady Salinity Intrusion in Estuaries.* La Houille Blanche **1/2**, 25–81.
31. Hayes, M.O. (1975). Morphology of sand accumulation in estuaries. In *Estuarine research*, Vol. II (L. Cronin, ed.), Academic Press, New York.
32. Holley, E.R. Harleman, D.R.F. and Fischer, H.B. (1970). Dispersion in homogeneous estuary flow. *Journal of the Hydraulics Division Proceedings*, ACSE **96**(HY8), 1691–1709.
33. Horrevoets, A.C. Savenije, H.H.G. Schuurman, J.N. and Graas, S. (2004). The influence of river discharge on tidal damping in alluvial estuaries. *Journal of Hydrology*, **294**, 213–228.
34. Hunt, J.N. (1964). Tidal oscillations in estuaries. *Geophysical Journal of Royal Astronomical Society*, **8**, 440–455.
35. Ippen, A.T. (1966a). Tidal dynamics in estuaries Part I: Estuaries of rectangular section. In *Estuary and Coastline Hydrodynamics*, (A.T. Ippen, ed.), McGraw-Hill, New York, USA, 493–521.
36. Ippen, A.T. (1966b). Salinity intrusion in estuaries. In *Estuary and Coastline Hydrodynamics*, (A.T. Ippen, ed.), McGraw-Hill, New York, USA, 598–629.
37. Ippen, A.T. and Harleman, D.R.F. (1961). *One-dimensional Analysis of Salinity Intrusion in Estuaries.* Technical Bulletin No. 5, Committee on Tidal Hydraulics, Corps of Engineers, U.S. Army, Vicksburg.
38. Ippen, A.T. and Harleman, D.R.F. (1966). Tidal Dynamics in Estuaries. In *Estuary and Coastline Hydrodynamics.* McGraw-Hill, New York, USA, 493–545.
39. Ippen, A.T. (1966). *Estuary and Coastline Hydrodynamics*, (A.T. Ippen, ed.), McGraw-Hill, New York, USA.
40. Jansen, P.Ph., van Bendegom, L., van den Berg, J., de Vries, M., and Zanen, A. (1979, 1983). *Principles of River Engineering, the non-tidal Alluvial River.* Pitman Publishing Ltd London.

41. Jay, D.A. (1991). Green's Law Revisited: Tidal long-wave propagation in channels with strong topography. *Journal of Geophysical Research*, **96**(C11), 20,585–20,598.
42. Jay, D.A. and Smith, J.D. (1990a). Residual circulation in shallow estuaries, 2. Weakly stratified and partially mixed, narrow estuaries. *Journal of Geophysical Research*, **95**(C1) 733–748.
43. Jay, D.A. and Smith, J.D. (1990b). Circulation, density structure and neap-spring transitions in the Columbia River Estuary. *Progress in Oceanography*, **25**, 81–112.
44. Jay, D.A. Geyer, W.R. Uncles, R.J. Vallino, J. Largier, J. and Boynton, W.R. (1997). A review of recent developments in estuarine scalar flux estimation. *Estuaries*, **20**(2), 262–280.
45. Kennedy, R.G. (1894). The prevention of silting in irrigation canals. *Minutes of the Proceedings of the Institute of Civil Engineers*, London, **119**, 281–290.
46. Kent, R.E. (1958). *Turbulent diffusion in a Sectionally Homogeneous Estuary*. Technical Report 16, Ref. 58-I. Chesapeake Bay Institute, John Hopkins University, Baltimore.
47. Ketchum, B.H. (1951). The exchanges of fresh and salt water in tidal estuaries. *Journal of Marine Research*, **10**(1), 18–38.
48. Kranenburg, C. (1985). *Alluviaal Estuarium, Commentaar Rapport H.H.G. Savenije*. personal handwritten communication, Vakgroep Vloeistofmechanica, Technische Hogeschool Delft.
49. Kranenburg, C. (1986). A time scale for long-term salt intrusion in well-mixed estuaries. *Journal of Physical Oceanography*, **16**, 1329–1331.
50. Lacey, G. (1930). Stable channels in alluvium. *Minutes of the Proceedings of the Institute of Civil Engineers*, London, **229**, 259–292.
51. Lacey, G. (1963). Discussion of Simons and Albertson (1960). Uniform water conveyance channels in alluvial material". *Transaction of the American Society of Civil Engineers*, **128(1)**, 65–167.
52. Lagrange, J. (1788). *Mécanique Analytique*. Desaínt, Paris.
53. Lamb, (1932). *art175. Hydrodynamics*. Cambridge University Press, Cambridge.
54. Lane, E.W. (1955). Design of stable channels. *ASCE Transactions*, **120**, 1234–1260.
55. Langbein, W.B. (1963). The hydraulic geometry of a shallow estuary. *Bulletin of International Association of Scientific Hydrology*, **8**, 84–94.
56. Lanzoni, S. and Seminara, G. (1998). On tide propagation in convergent estuaries. *Journal of Geophysical Research*, **103**(C13), 30,793–30,812.
57. Lanzoni, S. and Seminara, G. (2002). Long-term evolution and morphodynamic equilibrium of tidal channels. *Journal of Geophysical Research*, **107**(C1), 1–13, 3001; doi 10.102g/2000JC000468.
58. Leopold, L.B. and Maddock, T. (1953). *The Hydraulic Geometry of Stream Channels and some Physiographic Implications*. Professional Paper No. 252, U.S. Geological Survey, Washington D.C.
59. Leopold, L.B. Wolman, M.G. and Miller, J.P. (1964). *Fluvial Processes in Geomorphology*. Freeman, San Francisco.
60. Li, C. and O'Donnell, J. (1997). Tidally driven residual circulation in shallow estuaries with lateral depth variation. *Journal of Geophysical Research*, **102**(C13), 27,915–27,929.
61. Lindley, E.S. (1919). *Regime channels*. Proceedings of Punjab Engineering. Congress 7, pp. 63–74.
62. Manning, R. (1891). On the flow of water in open channels and pipes. *Transactions of the Institution of Civil engineers of Ireland*, Dublin, **20**, 161–207.
63. Mazure, J.P. (1937). *De berekening van getijden en stormvloeden op benedenrivieren*. Ph.D. Thesis, Delft University of Technology, Delft, The Netherlands.
64. McCarthy, R.K. (1993). Residual currents in tidally dominated, well-mixed estuaries. *Tellus*, *45A*, 325–340.

65. McDowell, D.M. and O'Connor, B.A. (1977). *Hydraulic Behaviour of Estuaries*. The Macmillan Press, London, UK.
66. Nichols, M.M. and Biggs, R.B. (1985). Estuaries. In *Coastal Sedimentary Environments* (R.A. Davis ed.), Springer-Verlag, New York, pp. 77–186.
67. O'Brien, M.P. (1931). Estuary tidal prisms related to entrance areas. *Civil Engineering*, **1(8)**, 738–739.
68. O'Kane, J.P. (1980). *Estuarine Water Quality Management*. Pitman Publishing Ltd, London, UK.
69. Okubo, A. (1967). The effect of shear in an oscillatory current on horizontal diffusion from an instantaneous source. *International Journal of Sea Research*, **6**, 213–224.
70. Pagès, J. and Citeau, J. (1990). Rainfall and Salinity of a Sahelian estuary between 1927 and 1987. *Journal of Hydrology*, **113**, 325–341.
71. Park, J.K. and James, A. (1990). Mass flux estimation and mass transport mechanism in estuaries. *Limnological Oceanography*, **35**(6), 1301–1313.
72. Pethick, J. (1984). *An Introduction to Coastal Geomorphology*. Edward Arnold Publishers, London.
73. Pillsbury, G. (1956). *Tidal Hydraulics*, First edition 1939. Corps of Engineers, Vicksburg, USA.
74. Ponce, V.M. and Simons, D.B. (1977). Shallow wave propagation in open channel flow. *Journal of the Hydraulics Division*, **103**(HY12), 1461–1476.
75. Prandle, D. (1981). Salinity intrusion in estuaries. *Journal of Physical Oceanography*, **11**, 1311–1324.
76. Prandle, D. (1985). On salinity regimes and the vertical structure of residual flows in narrow tidal estuaries. *Estuarine, Coastal and Shelf Science*, **20**, 615–635.
77. Prandle, D. (2003). Relationships between tidal dynamics and bathymetry in strongly convergent estuaries. *Journal of Physical Oceanography*, **33**, 2738–2750.
78. Preddy, W.S. (1954). The mixing and movement of water in the estuary of the Thames. *Journal of Marine Biological Association UK*, **33**(3), 645–662.
79. Rattray, M. Jr. and Dworski, J.G. (1980). Comparison of methods for analysis of the transverse and vertical circulation contributions to the longitudinal advective salt flux in estuaries. *Estuarine Coastal Marine Science*, **11**, 515–536.
80. Riggs, H.C. (1974). *Flash Flood Potential from Channel Measurements*. Symposium on Flash Floods IAHS, Publication, 112, pp. 52–56.
81. Rigter, B.P. (1973). Minimum length of salt intrusion in estuaries. *Journal of the Hydraulics Division Proceedings ASCE*, **99**, 1475–1496.
82. Risley, J.C. Guertin, D.P. and Fogel, M.M. (1993). Salinity intrusion forecasting system for Gambia river estuary. *Journal of Water Resources Planning and Management ASCE*, **119** (3), 339–352.
83. Rodriguez-Iturbe, I. and Rinaldo, A. (1997). *Fractal River Basins; Chance and Self-organisation*. Cambridge University Press, New York/Cambridge.
84. Rodriguez-Iturbe, I. and Rinaldo, A. (2001). *Fractal River Basins; Chance and Self-organisation* (first published 1997). Cambridge University Press, New York/Cambridge.
85. Savenije, H.H.G. (1986). A one-dimensional model for salinity intrusion in alluvial estuaries. *Journal of Hydrology*, **85**, 87–109.
86. Savenije, H.H.G. (1988). Influence of Rain and Evaporation on Salt Intrusion in Estuaries. *Journal of Hydraulic Engineering*, **114**(12), 1509–1524.
87. Savenije, H.H.G. (1989). Salt intrusion model for high-water slack, low-water slack and mean tide on spreadsheet. *Journal of Hydrology*, **107**, 9–18.
88. Savenije, H.H.G. (1992a). *Rapid Assessment Technique for Salt Intrusion in Alluvial Estuaries*. Ph.D. Thesis, IHE report series, no. 27, International Institute for Infrastructure, Hydraulics and Environment, Delft, The Netherlands.

89. Savenije, H.H.G. (1992b), Lagrangean solution of St. Venant's equations for an alluvial estuary. *Journal of Hydraulic Engineering*, **118**(8), 1153–1163.
90. Savenije, H.H.G. (1993a). Determination of estuary parameters on the basis of Lagrangian analysis. *Journal of Hydraulic Engineering*, **119**(5), 628–643.
91. Savenije, H.H.G. (1993b). Composition and driving mechanisms of longitudinal tidal average salinity dispersion in estuaries. *Journal of Hydrology*, **144**, 127–141.
92. Savenije, H.H.G. (1993c). Predictive model for salt intrusion in estuaries. *Journal of Hydrology*, **148**, 203–218.
93. Savenije, H.H.G. (1998). Analytical expression for tidal damping in alluvial estuaries. *Journal of Hydraulic Engineering*, **124**(6), 615–618.
94. Savenije, H.H.G. (2001a). A simple analytical expression to describe tidal damping or amplification. *Journal of Hydrology*, **243**, 205–215.
95. Savenije, H.H.G. (2001b). Equifinality, a blessing in disguise? HP Today Invited commentary, *Hydrological Processes*, **15**, 2835–2838.
96. Savenije, H.H.G. (2003). The width of a bankfull channel; Lacey's formula explained. *Journal of Hydrology*, **276**(1–4), 176–183.
97. Savenije, H.H.G. and Pagès, J. (1992). Hypersalinity, a dramatic change in the hydrology of Sahelian estuaries. *Journal of Hydrology*, **135**, 157–174.
98. Savenije, H.H.G. and Veling, E.J.M. (2005). The relation between tidal damping and wave celerity in estuaries. *Journal of Geophysical Research*, **110**, C04007, 1–10.
99. Schijf, J.B. and Schönfeld, J.C. (1953). *Theoretical Considerations on the Motion of Salt and Fresh Water*. Proceedings of Minnesota International Hydraulics Convention, Minneapolis, Minnesota 321–333.
100. Simons, D.B. and Albertson, M.L. (1960). Uniform water conveyance channels in alluvial material. *Journal of the Hydraulics Division ASCE*, **86**(5), 33–71.
101. Sivapalan, M. (2003). Process complexity at hillslope scale, process simplicity at the watershed scale: is there a connection? *Hydrological Processes*, **17**, 1037–1041.
102. Sivapalan, M. Bloeschl, G. Zhang, L. and Vertessy, R. (2003). Downward approach to hydrological prediction. *Hydrological Processes*, **17**, 2101–2111.
103. Smith, R. (1980). Buoyancy effects upon longitudinal dispersion in wide well-mixed estuaries. *Transaction of the Royal Philosophical Society* (London), **296**, 467–496.
104. Sobey, J. (2001). Evaluation of numerical models of flood and tide propagation in channels. *Journal of Hydraulic Engineering*, **127**(10), 805–824.
105. Stacey, M.T. Burau, J.R. and Monismith, S.G. (2001). Creation of residual flows in a partially stratified estuary. *Journal of Geophysical Research*, **106**(C8), 17,013–17,037.
106. Stevens, M.A. and Nordin, C.F. (1987). Critique of the regime theory for alluvial channels. *Journal of Hydraulic Engineering ASCE*, **113**(11), 1359–1380.
107. Stevens, M.A. (1989). Width of straight alluvial channels. *Journal of Hydraulic Engineering, ASCE*, **115**(3), 309–326.
108. Stigter, C. and Siemons, J. (1967). *Calculation of longitudinal salt distribution in estuaries as function of time*. Publ. 52. Delft Hydraulics Laboratory, The Netherlands.
109. Stommel, H. (1953). Computation of pollution in a vertically mixed estuary. *Sewage and Industrial Wastes*, **24**, 1065–1071.
110. Strickler, A. (1923). *Beitrage zur Frage der Geschwindigheidsformel und der Rauhigkeitszalen für Ströme, Kanäle und geschlossene Leitungen*. Mitteilungen des eidgenössischen Ambtes für Wasserwirtschaft 16, Bern, Switzerland.
111. Thatcher, M.L. and Harleman, D.R.F. (1972). *A Mathematical Model for the Prediction of Unsteady Salinity Intrusion in Estuaries*. R.M. Parsons Laboratory Report, No. 144, MIT, Cambridge, Massachusetts.

112. Turrell, W.R. Brown, J. and Simpson, J.H. (1996). Salt intrusion and secondary flow in shallow, well-mixed estuary. *Estuarine, Coastal and Shelf Science*, **42**, 153–169.
113. Uncles, R.J. and Stephens, J.A. (1996). Salt intrusion in the Tweed Estuary. *Estuarine, Coastal and Shelf Science*, **43**, 271–293.
114. Van Dam, G.C. and Schönfeld, (1967). *Experimental and Theoretical work in the field of Turbulent Diffusion Performed with Regard to The Netherlands' Estuaries and Coastal Regions of the North Sea.* Paper presented at General Assembly of IUGG, Berne, Switzerland.
115. Van de Kreeke, J. and Zimmerman, J.T.F. (1990). Gravitational circulation in well- and partially mixed estuaries, Chapter 14. In *Ocean Engineering Science*: The Sea, Vol. **19**, Part A. (B. le Méhanté and D.M. Hanes, eds), John Wiley & Sons Inc., New York.
116. Van den Burgh, P. (1972). *Ontwikkeling van een methode voor het voorspellen van zoutverdelingen in estuaria, kanalen en zeeen.* Rijkswaterstaat Rapport, 10–72.
117. Van Os, A.G. and Abraham, G. (1990). *Density Currents and Salt Intrusion.* Delft Hydraulics, International Institute for Hydraulic and Environmental Engineering, Lecture Notes.
118. Van Rijn, L. (1990). *Principles of Fluid Flow and Surface Waves in Rivers, Estuaries, Seas and Oceans.* Aqua Publications, Amsterdam.
119. Van Veen, J. (1937). *Getijstroomberekening met behulp van wetten analoog aan die van Ohm en Krichhoff.* De Ingenieur, no. 19.
120. Van Veen, J. (1950). *Eb- en vloedschaarsystemen in de Nederlandse getijwateren.* Tijdschrift Koninklijk Nederlands Aardrijkskundig Genootschap 67, 303–325.
121. West, J.R. and Broyd, T.W. (1981). Dispersion coefficients in estuaries. *Proceedings of the Institution of Civil Engineers*, **71**, 721–737.
122. West, J.R. and Mangat, J.S. (1986). The determination and prediction of longitudinal dispersion coefficients in a narrow, shallow estuary. *Estuarine, Coastal and Shelf Science* **22**, 161–181.
123. Whitham, G.B. (1974). *Linear and nonlinear waves.* John Wiley and Sons, New York, USA.
124. Wolanski, E. (1986). An evaporation driven salinity maximum zone in Australian tropical estuaries. *Estuarine, Coastal and Shelf Science*, **22**, 415–424.
125. Wright, L.D. Coleman, J.M. and Thom, B.G. (1973). Processes of channel development in a high tide range environment: Cambridge Gulf-Ord river delta, Western Australia. *Journal of Geology*, **81**, 15–41.
126. Wright, L.D. Coleman, J.M. and Thom, B.G. (1975). Sediment transport and deposition in a macrotidal river channel. In *Estuarine Research* (L.E. Cronin, ed.), Academic Press, New York.

Index

Printed and bound by CPI Group (UK) Ltd, Croydon, CR0 4YY
08/05/2025
01864806-0004